Morton L. Curtis

Matrix Groups

Springer-Verlag
New York Heidelberg Berlin

Universitext

Morton L. Curtis

Matrix Groups

Springer-Verlag
New York   Heidelberg   Berlin

Dr. Morton L. Curtis
Department of Mathematics
Rice University
Houston, Texas 77001
USA

AMS Classifications: 20-01, 20G99, 22E20

Library of Congress Cataloging in Publication Data

Curtis, Morton Landers, 1921–
    Matrix groups.

    (Universitext)
    Includes index.
    1. Matrix groups. I. Title.
QA171.C87   512′.2   79-23523

© 1979 by Springer-Verlag New York Inc.

Printed in the United States of America.

9 8 7 6 5 4 3 2 1

ISBN 0-387-90462-X   Springer-Verlag   New York
ISBN 3-540-90462-X   Springer-Verlag   Berlin   Heidelberg

To my teacher and friend
Raymond Louis Wilder
this book is affectionately dedicated.

Introduction

These notes were developed from a course taught at Rice University in the spring of 1976 and again at the University of Hawaii in the spring of 1977. It is assumed that the students know some linear algebra and a little about differentiation of vector-valued functions. The idea is to introduce students to some of the concepts of Lie group theory-- all done at the concrete level of matrix groups. As much as we could, we motivated developments as a means of deciding when two matrix groups (with different definitions) are isomorphic.

In Chapter I "group" is defined and examples are given; homomorphism and isomorphism are defined. For a field k, $M_n(k)$ denotes the algebra of $n \times n$ matrices over k. We recall that $A \in M_n(k)$ has an inverse if and only if $\det A \neq 0$, and define the general linear group $GL(n,k)$. We construct the skew-field $\mathbb{H}$ of quaternions and note that for $A \in M_n(\mathbb{H})$ to operate linearly on $\mathbb{H}^n$ we must operate on the right (since we multiply a vector by a scalar on the left). So we use row vectors for $\mathbb{R}^n$, $\mathbb{C}^n$, $\mathbb{H}^n$ and write xA for the row vector obtained by matrix multiplication. We get a complex-valued determinant function on $M_n(\mathbb{H})$ such that $\det A \neq 0$ guarantees that A has an inverse.

Chapter II introduces conjugation on $\mathbb{R} \subset \mathbb{C} \subset \mathbb{H}$ and then an inner product $\langle \, , \, \rangle$. Basic properties of $\langle \, , \, \rangle$ are given and then for $k \in \{\mathbb{R}, \mathbb{C}, \mathbb{H}\}$ we define the orthogonal group

$$\mathbb{O}(n,k) = \{A \in M_n(k) \mid \langle xA, yA \rangle = \langle x,y \rangle \text{ for all } x,y \in k^n\} \quad .$$

$\mathfrak{G}(n,R)$ is written $\mathfrak{G}(n)$ and called the _orthogonal_ group.
$\mathfrak{G}(n,C)$ is written $U(n)$ and called the _unitary_ group. $\mathfrak{G}(n,\mathbb{H})$ is
written $Sp(n)$ and called the _symplectic_ group. If $A \in \mathfrak{G}(n)$ then
$\det A \in \{1,-1\}$ and the subgroup with $\det = 1$ is denoted by $SO(n)$
and called the _special_ _orthogonal_ _group_. If $A \in U(n)$ then $\det A$
is a complex number of unit length. The subgroup with $\det = 1$ is
denoted by $SU(n)$ and called the _special_ _unitary_ _group_. As a first
example of a matrix group isomorphism we show that $Sp(1) \cong SU(2)$.

In Chapter III we define the first invariant (i.e., something
unchanged by an isomorphism) of a matrix group; namely, its _dimension_.
A tangent vector to a matrix group G is $\gamma'(0)$ for some differentiable curve γ in G with $\gamma(0) = I$. The set T_G of all tangent
vectors is shown to be a vector space, a real subspace of $M_n(k)$
$(k \in \{R,C,\mathbb{H}\}$). The dimension of T_G (as a real vector space) is
the dimension of G . Smooth homomorphisms are defined and shown
to induce linear maps of tangent spaces. Then dimension is seen to
be an invariant.

In order, in Chapter IV, to calculate the dimensions of our matrix
groups we develop the exponential map $\exp: M_n(k) \to GL(n,k)$ and
the logarithm, $\log: U \to M_n(k)$ where U in some neighborhood of I
in $GL(n,k)$. We have that $\exp$ and $\log$ are inverses, $\exp: V \to U$
and $\log: U \to V$ where V is a neighborhood of 0 on $M_n(k)$ and
U is a neighborhood of I in $M_n(k)$ (actually $U \subset GL(n,k)$). One-
parameter subgroups are defined and proved to be determined by their
derivatives at 0 . It follows that T_G can be taken to be all
derivatives of one-parameter subgroups. Lie algebras are defined and

we see that each T_G is a Lie algebra. Finally, we then calculate the dimensions of SO(n) , U(n) , SU(n) and Sp(n) .

In Chapter V we consider the very specific question of whether Sp(1) and SO(3) are isomorphic. We get a surjective homomorphism ρ: Sp(1) → SO(3) with kernel = {1,-1} . Then we define the center of a group, show it is an invariant and then calculate Center Sp(1) = {1,-1} and Center SO(3) = {I} , proving that Sp(1) ≠ SO(3) . We define quotient groups and then note that we get new groups $\frac{G}{\text{center}}$ whenever a matrix group has nontrivial center.

In Chapter VI we do some topology which is needed in other parts of the text. All of our matrix groups are in some euclidean space and we just do topology of subsets of euclidean spaces. We give some basic results about continuity of functions, connected sets and compact sets. The proof that continuous functions preserve compactness relegated to an appendix. We consider countable bases for open sets since this is needed later in our study of maximal tori in matrix groups. Finally, there is a short section on manifolds.

Chapters VII, VIII, and IX are devoted to studying maximal tori in our matrix groups. We describe certain specific maximal tori. We prove that any two maximal tori are conjugate and that, if G is connected, then these conjugates cover G . At this stage we then know the dimension, center and rank of all of our matrix groups and these suffice to settle our original question as to which of these groups are isomorphic. At the end of Chapter IX we discuss simple groups and covering groups. The only new groups which arise are the double covers of SO(n) (n = 3,4,...). This leads to the question:

Is the double cover of $SO(2n+1)$ isomorphic with $Sp(n)$?

In Chapter X we construct the double cover $Spin(n)$ of $SO(n)$ using Clifford algebras. We show that $Spin(1) \simeq S^0$, $Spin(2) \simeq S^1$, $Spin(3) \simeq Sp(1)$ $(= S^3)$ and $Spin(4) \simeq Sp(1) \times Sp(1)$. Finally we show that

$$Spin(5) \simeq Sp(2) \quad \text{and} \quad Spin(6) \simeq SU(4) .$$

In Chapter XI we finish our job by showing that $Sp(n) \neq Spin(2n+1)$ for any $n > 2$. This is done by looking at normalizers of maximal tori and resulting Weyl groups. If the normalizer is the semidirect product of the torus and the Weyl group, we say that the normalizer splits. If $Spin(2n+1)$ and $Sp(n)$ were isomorphic we would have

$$\frac{Spin(2n+1)}{center} \simeq \frac{Sp(n)}{center} .$$

So our result is a consequence of the following three results.

(*) The normalizer in $Sp(n)$ does not split for any n .

(+) The normalizer in $\frac{Sp(n)}{center}$ splits $\Leftrightarrow n \in \{1,2\}$.

(T) The normalizer in $\frac{Spin(2n+1)}{center}$ ($= SO(2n+1)$) splits for $n = 1,2,3,\ldots$.

Finally, in Chapter XII we give a brief introduction to abstract Lie groups.

CONTENTS

Chapter 1
General Linear Groups

A. Groups

Before we can discuss matrix groups we need to talk a little about groups in general. If X and Y are sets, their <u>Cartesian product</u> $X \times Y$ is defined to be the set of all ordered pairs (x,y) with $x \in X$ and $y \in Y$. A convenient notation for describing this set of all ordered pairs is

$$X \times Y = \{(x,y) \mid x \in X \quad \text{and} \quad y \in Y\} ,$$

the curly brackets being read as "the set of all" and the vertical bar as "such that."

By a <u>binary operation</u> ρ on a set S we mean a function

$$\rho : S \times S \to S ;$$

i.e., for an ordered pair (s_1, s_2) of elements of S, ρ assigns another element of S which we write as $\rho(s_1, s_2)$. For example, the set $N = \{1,2,3,\ldots\}$ of natural numbers has two well-known binary operations on it. Addition sends the ordered pair (a,b) of natural numbers to the natural number $a + b$. Multiplication sends the ordered pair (a,b) to ab.

Definition: A group G is a set G along with a binary operation

$$\phi : G \times G \to G$$

satisfying certain properties. To state these properties it is convenient to adopt a simple notation--for $\phi(a,b)$ we just write ab .

Required properties of the operation:

(i) The operation is associative. This means that for any a,b,c $\in$ G we have

$$(ab)c = a(bc) .$$

(If we had maintained the $\phi(a,b)$ notation this would read $\phi(\phi(a,b),c) = \phi(a,\phi(b,c))$.

(ii) There exists an identity element e of G . This means that for any a $\in$ G we have ea = ae = a .

(iii) Inverses exist. This means that for any a $\in$ G there is an element $a^{-1} \in$ G such that $aa^{-1} = a^{-1}a = e$.

Note that properties (ii) and (iii) leave open the possibilities that there may be more than one identity element and that an element may have more than one inverse. But neither of these can happen.

Proposition 1: A group G has exactly one identity element and each a $\in$ G has exactly one inverse.

Proof: Suppose e and f are identity elements of G . Then fe = e   since   f   is an identity element, and

fe = f   since   e   is an identity element .

Suppose both   b   and   c   are inverses of   a .   Then

$$b = eb = (ca)b = c(ab) = ce = c .$$

Examples

(1)   The set $Z = \{\ldots,-2,-1,0,1,2,\ldots\}$ of integers is a group under addition.   0   is the identity and the inverse of   a   is   -a .

(2) Z is not a group under multiplication.   The operation is associative and   1   is the identity.   But, for example, there is no inverse for   2 .

(3)   The set Q of rational numbers is a group under addition.

(4)   The set $Q - (0)$ (i.e., all nonzero rationals) is a group under multiplication.

(5) $R^{+} = \{x \in R | x > 0\}$ is the set of all positive real numbers.   It forms a group under multiplication.

(6) R^{n} = the set of all ordered n-tuples of real numbers is a group under the following operation: if

$$x = (x_1, x_2, \ldots, x_n) \quad \text{and}$$

$$y = (y_1, y_2, \ldots, y_n), \quad \text{then}$$

$$x + y = (x_1 + y_1, x_2 + y_2, \ldots, x_n + y_n) .$$

The identity   is

$$\textcircled{0} = (0,0,\ldots,0)$$

and the inverse of x is $(-x_1,-x_2,\ldots,-x_n)$.

(7) Let $S = \{a,b,c\}$; i.e., S is a set with three elements which we denote by a,b,c . Let G be the set of all one-to-one maps (functions) of S onto S . For example $f : S \to S$ given by $f(a) = b$, $f(b) = c$, $f(c) = a$ is one element of G . We define an operation on G as follows: if $f,g \in G$ we let

$$f \circ g : S \to S$$

be defined by $(f \circ g)(a) = f(g(a))$, $(f \circ g(b) = f(g(b))$, $(f \circ g)(c)$ $= f(g(c))$, i.e., $f \circ g$ means first apply g to S and then apply f. Let $i : S \to S$ be the identity element $(i(a) = a, i(b) = b, i(c) = c)$. Then this is the identity element for G for this operation. Then the usual inverse of $f \in G$ is the inverse for f relative to this operation. Thus G is a group. It is called the symmetric group on $\{a,b,c\}$ (or just the symmetric group on three elements).

Definition: A group G is abelian if for every $a,b \in G$ we have $ab = ba$.

In the examples above, (1), (3), (4), (5), and (6) are abelian groups, but the symmetric group on three elements is not abelian. (Exercise.)

The kind of functions (mapping one group to another) of interest to us are those which "preserve" the operations--these are called homomorphisms.

Definition: Let G and H be groups. A function $\sigma : G \to H$ is a homomorphism if for every a,b in G we have

$$\sigma(ab) = \sigma(a)\sigma(b) .$$

What this means is that we can first multiply a and b (using the operation in G) and then map the result by σ , or we can map a and b into H by σ and multiply there--with the same result.

Proposition 2: A homomorphism $\sigma : G \to H$ sends identity to identity and inverses to inverses.

Proof: Let e,e' be the identities in G,H . We have $\sigma(e) = \sigma(ee) = \sigma(e)\sigma(e)$ and $\sigma(e)$ has an inverse, call it h , in H . So

$$e' = h\sigma(e) = h\sigma(e)\sigma(e) = \sigma(e) .$$

For $a \in G$ we have

$$\sigma(a)\sigma(a^{-1}) = \sigma(aa^{-1}) = \sigma(e) = e' ,$$

showing that $\sigma(a^{-1}) = (\sigma(a))^{-1}$.

A homomorphism is surjective (or onto) if $\sigma(G) = H$. If we define $\sigma : R \to R^2$ (R = additive group of reals, R^2 as in example (6)) by $\sigma(x) = (x,x)$, then σ is a homomorphism but is not surjective because $\sigma(R)$ is just the diagonal line in R^2 . But $\rho : R^2 \to R$ defined by $\rho(x,y) = x$ is a surjective homomorphism.

A homomorphism $\sigma : G \to H$ is injective if $\sigma(a) = \sigma(b)$ always implies a = b ; i.e., no two elements go to the same place. Some-times this is called one-to-one-into, but we won't do that. For example, the map $\sigma : R \to R$ $(\sigma(x) = (x,x))$ is injective, and the map

$\rho : R^2 \to R$ $(_0(x,y) = x)$ is not injective.

A homomorphism which is both injective and surjective is called an _isomorphism_. From an abstract point of view, two groups which are isomorphic are "really" the same group--even if they were defined in strikingly different manners. There is a classic example of this.

Let   R   be the additive group of all real numbers and let R^+ (see Example 5) be the multiplicative group of all positive real numbers. Let   a   be any real number greater than   1 .   Define

$$\sigma : R \to R^+$$

by

$$\sigma(x) = a^x .$$

Then σ is a homomorphism

$$\sigma(x + y) = a^{x+y} = a^x a^y = \sigma(x)\sigma(y) .$$

Also, σ is injective. For, suppose $\sigma(x) = \sigma(y)$.   This means $a^x = a^y$ and so $a^{-y}a^x = a^{-y}a^y = 1$ and $a^{x-y} = 1$ which implies $x - y = 0$ or $x = y$.   Also, σ is surjective. For, if   y   is any positive real number $x = \log_a y$ has the property that $a^x = y$. Thus these two groups are isomorphic--not only that, but there are lots of isomorphisms.

We conclude this section with a simple, but important, remark. A priori it looks difficult to see if a homomorphism $\sigma : G \to H$ is injective. Do we really have to check all pairs   a,b   in   G   to see if $\sigma(a) = \sigma(b)$?   Fortunately not.

$$\sigma \text{ is injective} \Leftrightarrow \sigma^{-1}(e') = e .$$

$$\sigma(a) = \sigma(b) \Leftrightarrow \sigma(a)\sigma(b)^{-1} = e' \Leftrightarrow \sigma(ab^{-1}) = e'$$

and

$$ab^{-1} = e \Leftrightarrow a = b \ .$$

B. Fields, Quaternions

Definition: A field k is a set that has operations of addition and multiplication satisfying certain requirements:

(i) multiplication distributes over addition;

$$a(b + c) = ab + ac \ ;$$

(ii) k is an abelian group, with identity written as 0 , under addition.

(iii) k - (0) is an abelian group under multiplication.

Examples. The rationals $\mathbb{Q}$ and the reals $\mathbb{R}$ are fields. We can make $\mathbb{R}^2$ into a field $\mathbb{C}$ (the complex numbers) as follows. If (x_1,x_2) and (y_1,y_2) are two ordered pairs of real numbers, we define $(x_1,x_2) + (y_1,y_2) = (x_1 + y_1, x_2 + y_2)$ and we have seen that this operation makes $\mathbb{R}^2$ into an abelian group. Suppose for multiplication we try

$$(x_1,x_2)(y_1,y_2) = (x_1 y_1, x_2 y_2)$$

(surely the most obvious thing). Then we would have

$$(1,0)(0,1) = (0,0) \ .$$

Now (0,0) is the additive identity or "zero" and we would have two nonzero elements of R^2 with a zero product. The result could not be a field because:

Proposition 3: In a field k if $a \neq 0$ and $b \neq 0$, then

$$ab \neq 0 .$$

Proof: If $a \neq 0$ then $a \in k - (0)$ which by (iii) is required to be a group under multiplication. Thus there is an a^{-1} in $k - (0)$ such that $a^{-1}a = 1$ (the multiplicative identity). Thus if $ab = 0$ we have

$$a^{-1}(ab) = (a^{-1})(0) = 0$$

but $a^{-1}(ab) = (a^{-1}a)b = 1b = 0$ so $b = 0$.

The statement of Proposition 3 is equivalent to the statement that a field has "no divisors of zero."

So how do we make R^2 into a field? Our most naive attempt failed flat. Well, what turns out to work is

$$(a,b)(c,d) = (ac - bd, ad + bc) .$$

We must first verify that this distributes over addition.

$$(a,b)((c,d) + (e,f)) = (a,b)((c + e, d + f))$$

$$= (a(c + e) - b(d + f), a(d + f) + b(c + e)) .$$

This should equal $(a,b)(c,d) + (a,b)(e,f)$. This latter equals $(ac - bd, ad + bf) + (ad - bf, af + be)$ and we easily check that these are equal. Next we need to see that if $(a,b) \neq (0,0)$ then it has a multiplicative inverse. Well, $(a,b) \neq (0,0) \Leftrightarrow a^2 + b^2 \neq 0$; in which case, we need to find a multiplicative inverse for (a,b) . The multiplicative identity clearly is $(1,0)$ and

$$(a,b)(\frac{a}{a^2 + b^2} , \frac{-b}{a^2 + b^2}) = (1,0) ,$$

as you can readily verify. Thus we have made R^2 into a field which we denote by C and call the complex numbers.

You may know that there is a simple mnemonic device for remembering multiplication in C . Write $(a,b) = a + ib$ or $a + bi$ and treat these as polynomials in i with the side condition that $i^2 = -1$. Thus

$$(a + ib)(c + id) = ac + aid + ibc + ibid$$

$$= ac + iad + ibc + i^2 bd$$

$$= (ac - bd) + i(ad + bc) .$$

We can consider R to be a subfield of C (i.e., a subset which becomes a field using the operations in the larger set) by letting

$$x \in R \quad \text{be} \quad x + i0 .$$

Then if $x,y \in R$ we have

$$x + y = x + i0 + y + i0 = (x + y) + i0$$

$$xy = (x + i0)(Y + i0) = (xy) + i0 .$$

So we have taken the field R as all $(x,0)$ in R^2 and extended the operations in R to R^2 to get a field.

This strongly suggests that we try to extend the field on R^2 to a field on R^3. Now for the bad news.

<u>Proposition</u> 4: <u>The operations on</u> ℂ <u>cannot be extended to make</u> R^3 <u>into a field.</u>

<u>Proof</u>: Take basis vectors 1,i,j so that any element of R^3 can be written uniquely as $a+ib+jc$ with $a,b,c \in R$.

So if we are to have a multiplication extending that of ℂ we must have $ij = a+ib+jc$ for some three real numbers a,b,c. But then $i(ij) = ia+i^2b+ijc$; so

$$-j = ia - b + ijc$$

$$-j = ia - b + (a + ib + jc)c$$

$$-j = (a - b) + i(a + b) + jc^2 .$$

This implies $c^2 = -1$, contradicting $c \in R$.

The main thrust of this proof is that if we insist that the product ij be in R^3 we get into trouble. Maybe if we had one more dimension it would work. This is <u>almost</u> true; we can define a multiplcation on R^4 which satisfies conditions (i) and (ii) for a field but (iii) must be replaced by (iii)' $k - (0)$ is a group under multiplication--it is <u>not an abelian group</u>. We will just describe how this can be done. You may be interested in reading "Hamilton's discovery of the quaternions" by B. L. van der Waerden in the Mathematics Magazine (vol. 49, #5, (1976)). We take a basis 1,i,j,k for R^4

and define

	1	i	j	k
1	1	i	j	k
i	i	-1	k	-j
j	j	-k	-1	i
k	k	j	-i	-1

Thus 1 acts as identity, $ij = k$, $ji = -k$, etc.

This tells us how to multiply quadruples of real numbers:

$$(a + ib + jc + kd)(x + iy + jz + kw) = (ax - by - cz - dw)$$

$$+ i(ay + bx + cw - dz) + j(az + cx + dy - bw)$$

$$+ k(aw + bx + bz - cy) .$$

R^4 with this multiplication is called the <u>quaternions</u>. It is easy
to verify that this does extend the multiplication in C by taking
$c = 0 = d$ and $z = 0 = w$ in the formula above. The modified field
axioms (i), (ii), (iii)' are readily verified except for showing
that every nonzero quaternion has an inverse. But if

$$q = a + ib + jc + bd$$

is not the zero $(0 + i0 + j0 + k0)$ then $a^2 + b^2 + c^2 + d^2 \neq 0$ and
we set

$$q^{-1} = \frac{a - ib - jc - kd}{a^2+b^2+c^2+d^2} ,$$

and readily verify that $qq^{-1} = 1 = q^{-1}q$.

There are certain constructions we want to make for R and C
and the quaternions (which we denote by $\mathbb{H}$), so we will write

$$k \in \{ \mathbb{R}, \mathbb{C}, \mathbb{H} \} \ .$$

C.   Vectors and Matrices

For $k \in \{ \mathbb{R}, \mathbb{C}, \mathbb{H} \}$ let k^n be the set of all ordered n-tuples of elements of k . Define addition on k^n by

$$x = (x_1, \ldots, x_n) \qquad y = (y_1, \ldots, y_n)$$

$$x + y = (x_1 + y_1, \ldots, x_n + y_n) \ .$$

This makes k^n into an abelian group with identity $\mathbb{O} = (0, \ldots, 0)$. For $c \in k$ we define

$$cx = (cx_1, \ldots, cx_n)$$

and this makes k^n into a vector space over k (for $k = \mathbb{H}$ we must relax the usual definition which insists that k be a field).

Definition: A map $k^n \overset{\phi}{\to} k^n$ is linear if it respects linear combinations; i.e., if $c, d \in k$ and $x, y \in k^n$ then

$$(*) \qquad\qquad \phi(cx + dy) = c\phi(x) + d\phi(y) \ .$$

In particular, $\phi(x + y) = \phi(x) + \phi(y)$, so that a linear map is a homomorphism of the additive group of k^n . Also

$$\phi(cx) = c\phi(x) \ ,$$

and these two conditions together are equivalent to $(*)$.

Proposition 5: If $k^n \overset{\phi}{\to} k^n \overset{\psi}{\to} k^n$ are both linear, then so is $\psi \circ \phi$.

Proof: $(\psi \circ \phi)(cx + dy) = \psi(c\phi(x) + d\phi(y))$

$$= c(\psi \circ \phi)(x) + d(\psi \circ \phi)(y) .$$

Definition: $M_n(k)$ is the set of all $n \times n$ matrices with elements from k .

If $M \in M_n(K)$, $M = (m_{ij})$ $(m_{ij} \in k)$, we can define a linear map $\phi(M)$ by

$$\phi(M)(x_1, \ldots, x_n) = (x_1, \ldots, x_n)(m_{ij})$$

where matrix multiplication is indicated on the right; i.e., we are multiplying a $1 \times n$ matrix with an $n \times n$ matrix to give a $1 \times n$ matrix. This is easily seen to be linear.

$$\phi(M)(cx + dy) = (cx + dy) (m_{ij})$$

$$= c(x_1, \ldots, x_n)(m_{ij}) + d(y_1, \ldots, y_n)(m_{ij}) .$$

We use row vectors instead of column vectors because we no longer have a choice when $k = \mathbb{H}$. We made $\mathbb{H}^n$ into a vector space by defining scalar multiplication on the left,

$$c(x_1, \ldots, x_n) = (cx_1, \ldots, cx_n)$$

and this is not the same as $(x_1 c, \ldots, x_n c)$ in general. If we use column vectors and multiply by matrices on the left we do not always get linear maps. For $q, c, d \in \mathbb{H}$ and $x, y \in \mathbb{H}^n$ consider

$$\begin{pmatrix} q & & \\ & q & \\ & & \ddots & \\ & & & q \end{pmatrix} \begin{pmatrix} cx_1 + dy_1 \\ \vdots \\ cx_n + dy_n \end{pmatrix} = \begin{pmatrix} qcx_1 + qdy_1 \\ \vdots \\ qcx_n + qdy_n \end{pmatrix}$$

and we certainly can't expect this to equal

$$c \begin{pmatrix} qx_1 \\ \vdots \\ qx_n \end{pmatrix} + d \begin{pmatrix} qy_1 \\ \vdots \\ qy_n \end{pmatrix} \ .$$

(Take $n = 1$, $x = 1$, $y = 1$, $d = 0$, $c = i$ and $q = j$.)

Conversely, given a linear map $\phi : k^n \to k^n$, it is easy to find an $n \times n$ matrix M such that $\phi = \phi(M)$ (and it will clearly be unique). The first row of M is the n-tuple $\phi(1,0,\ldots,0)$, the second row of M is $\phi(0,1,0,\ldots,0)$, etc.

Note that if the matrix A gives the linear map ϕ and the matrix B gives the linear map ψ then AB gives $\psi \circ \phi$. A linear map ϕ is an <u>isomorphism</u> if it is injective and surjective (same definitions as for group homomorphisms). Then ϕ^{-1} is also a linear isomorphism and $\phi \circ \phi^{-1} = $ identity map $= \phi^{-1} \circ \phi$. For the corresponding matrices this means that $M(\phi^{-1})M(\phi) = I = M(\phi)M(\phi^{-1})$ so that $M(\phi^{-1})$ is a 2-sided inverse for $M(\phi)$. So <u>if</u> A^{-1} <u>is a left inverse for</u> A, <u>then it is also a right inverse for</u> A.

We make the set $M_n(k)$ into a vector space in a fairly obvious way:

(i) If $A = (a_{ij})$ and $B = (b_{ij})$, then

$$A + B = (a_{ij} + b_{ij}) \ ;$$

(ii) If $A = (a_{ij})$ and $c \in k$, then

$$cA = (ca_{ij}) .$$

This is really no different from the way we made k^n into a vector space, but we are now working with n^2-tuples. However, there is nothing to be gained by writing n^2 elements in a line instead of an n x n array.

But $M_n(k)$ is not just a vector space. It also has a multi-plication which distributes over addition (on either side).

$$A(B + C) = AB + AC$$

$$(B + C)A = BA + CA .$$

Such a system is called an _algebra_. When we use the word algebra we will always mean one with a two-sided multiplicative identity. For $M_n(k)$,

$$I = \begin{pmatrix} 1 & & \\ & 1 & & \bigcirc \\ & & \ddots & \\ & \bigcirc & & \ddots & \\ & & & & 1 \end{pmatrix}$$

is the multiplicative identity.

D. General Linear Groups

Definition: If G is an algebra, $x \in G$ is a _unit_ if there exists some $y \in G$ such that $xy = 1 = yx$, i.e., if it has a multi-plicative inverse.

Proposition 6: __If G is an algebra with an associative__

multiplication <u>and</u> U ⊂ G <u>is the set of units in</u> G , <u>then</u> U <u>is a group</u> <u>under</u> <u>multiplication</u>.

Proof: The operation is associative, there is an identity element 1 and every element has an inverse.

Definition: The group of units in the algebra $M_n(R)$ is denoted by GL(n,R) , in $M_n(C)$ by GL(n,C) and in $M_n(\mathbb{H})$ by GL(n,$\mathbb{H}$) . These are the general linear groups.

Note that: A ∈ $M_n(k)$ is a unit ⟺ A represents an isomorphism of k^n .

Definition: If G is a group and H is a subset of G , then H is a subgroup of G if the operation on G makes H into a group.

Proposition 7: H <u>is a subgroup of the group</u> G <u>if</u> (H ⊂ G <u>and</u>)

(i)   x,y ∈ H ⟹ xy ∈ H ,

(ii) <u>id. el. is in</u> H ,

(iii) x ∈ H ⟹ x^{-1} ∈ H .

Proof: (Exercise.) The subject of this course is the study of subgroups of these general linear groups.

A 1 × 1 matrix over k is just an element of k and matrix multiplication of two is just multiplication in k . So we see that

$$GL(1,R) = R - (0)$$
$$GL(1,C) = C - (0)$$
$$GL(1,\mathbb{H}) = \mathbb{H} - (0)$$

because all nonzero elements are units. $GL(2,R)$ is the set of units in the vector space $M_2(R)$ of dimension 4 . So

$$GL(2,R) = \{(\begin{smallmatrix} a & b \\ c & d \end{smallmatrix}) \mid \begin{smallmatrix} a,b,c,d \in R \\ ad - bc \neq 0 \end{smallmatrix}\} ,$$

i.e., all points in 4-space not on the set where $ad = bc$.

For R and C we have determinants defined on $M_n(R)$ and $M_n(C)$ and from linear algebra we know that

$$GL(n,R) = \{A \in M_n(R) \mid \det A \neq 0\}$$

$$GL(n,C) = \{A \in M_n(C) \mid \det A \neq 0\} .$$

Suppose we define a "determinant" on $M_2(\mathbb{H})$ by

$$\det(\begin{smallmatrix} \alpha & \beta \\ \gamma & \delta \end{smallmatrix}) = \alpha\delta - \beta\gamma .$$

Then $\det(\begin{smallmatrix} i & j \\ i & j \end{smallmatrix}) = k - (-k) = 2k \neq 0$, but this matrix cannot be a unit or the corresponding linear map would be an isomorphism, whereas

$$(j,-j)(\begin{smallmatrix} i & j \\ i & j \end{smallmatrix}) = (0,0)$$

and the map is not injective. Similar definitions give similar problems, but we can define a complex-valued determinant with the desired property: namely, $A \in M_n(\mathbb{H})$ has an inverse if and only if this determinant is nonzero.

Proposition 8: Let $\phi : G \to H$ be a homomorphism of groups. Then $\phi(G)$ is a subgroup of H.

Proof: $\phi(id) = id$ so that $\phi(G)$ contains the identity element of H. If $x,y \in \phi(G)$ there exist $a,b \in G$ such that $\phi(a) = x$, $\phi(b) = y$. Then

$$xy = \phi(a)\phi(b) = \phi(ab) \in \phi(G) \ .$$

Finally, suppose $x \in \phi(G)$. Then $x = \phi(a)$ and so $x^{-1} = \phi(a^{-1})$ $\in \phi(G)$. So $\phi(G)$ is a subgroup of H .

If $\phi : G \to H$ is an injective homomorphism, then ϕ is an isomorphism of G onto the subgroup $\phi(G)$ of H , so we can then consider G as a subgroup of H . We are going to construct an injective homomorphism

$$\Psi : GL(n,\mathbb{H}) \to GL(2n,\mathbb{C}) \ ,$$

and then for $A \in GL(n,\mathbb{H})$ we will assign as the determinant of A the determinant of $\Psi(A)$.

We begin with

$$\psi : \mathbb{H} \to M_2(\mathbb{C})$$

defined by

$$\psi(x+iy+jz+kd) = \begin{pmatrix} x+iy & -z-iw \\ z-iw & x-iy \end{pmatrix} \ .$$

Lemma 9:   (i) $\psi(\alpha+\beta) = \psi(\alpha) + \psi(\beta)$

(ii) $\psi(\alpha\beta) = \psi(\alpha)\psi(\beta)$

(iii) ψ is injective .

Proof: (i) is trivial and (ii) is a routine, but somewhat tedious computation, and (iii) is trivial.

Next, for $A \in M_n(\mathbb{H})$ we set

$$\psi(A) = (\psi(a_{ij}))$$

i.e. $\psi(A)$ is the complex $2n \times 2n$ matrix whose 2×2 block in the ij position is $\psi(a_{ij})$.

Lemma 10: $\psi(AB) = \psi(A)\psi(B)$.

Proof: Let $A = (\alpha_{\mu\nu})$, $B = (\beta_{\mu\nu})$. Then

$$(AB)_{ij} = \alpha_{i1}\beta_{1j} + \cdots + \alpha_{in}\beta_{nj} . \quad \text{By Lemma 9}$$

$$(\psi(AB)_{ij}) = \psi(\alpha_{i1})\psi(\beta_{1j}) + \cdots + \psi(\alpha_{in})\psi(\beta_{nj})$$

and this is just the ij entry in $\psi(A)\psi(B)$.

Now let $A \in GL(n,\mathbb{H})$ so that there exists $A^{-1} \in GL(n,\mathbb{H})$ with $AA^{-1} = I = A^{-1}A$. Then $\psi(A)$ has $\psi(A^{-1}) = (\psi(A))^{-1}$ so that $\psi(A)$ is nonsingular and thus $\det \psi(A) \neq 0$.

Conversely, suppose $\det \psi(A) \neq 0$. Then $(\psi(A))^{-1}$ exists and since $\psi(GL(n,\mathbb{H}))$ is a subgroup of $GL(2n,\mathbb{C})$ we have that $(\psi(A))^{-1} \in \psi(GL(n,\mathbb{H}))$. Thus $\exists A^{-1} \in GL(n,\mathbb{H})$ such that $\psi(A^{-1}) = (\psi(A))^{-1}$. Then

$$\psi(AA^{-1}) = I$$

and ψ is injective so $AA^{-1} = I$. Thus A is nonsingular.

E. Exercises

1. Let $\phi : G \to H$ be a homomorphism of groups. The kernel of ϕ is defined to be

$$\ker \phi = \{x \in G \mid \phi(x) = \text{identity of } H\} \quad .$$

Show that $\ker \phi$ is a subgroup of G .

2. A subgroup W of a group G is <u>normal</u> if for each $x \in G$ we have

$$xWx^{-1} = W \quad .$$

Show that $\ker \phi$ (Exercise 1) is a normal subgroup of G .

3. The <u>center</u> C of a group G is defined by

$$C = \{y \in G \mid xy = yx \quad \text{for all} \quad x \in G\} \quad .$$

Show that C is a normal subgroup of G .

4. Let S be a nonempty subset in a group G . Define the <u>centralizer</u> $C(S)$ of S by

$$C(S) = \{x \in G \mid xs = sx \quad \text{for all} \quad x \in S\} \quad .$$

Show that $C(S)$ is a subgroup of G .

5. Let S be a nonempty set in a group G . Define

$$N(S) = \{x \in G \mid xSx^{-1} = S\}$$

and call $N(S)$ the <u>normalizer</u> of S . Show that $C(S) \subset N(S)$ and that $N(S)$ is a subgroup of G . Show that if S is a subgroup of G , then $S \subset N(S)$ and S is a normal subgroup of $N(S)$.

6. Show that if $\{H_\alpha \mid \alpha \in A\}$ is any collection of subgroups of G , then their intersection is also a subgroup of G . If W is any

subset of G , by the <u>subgroup generated by</u> W we mean the inter-section of all subgroups of G which contain W . Show that this is the smallest subgroup of G which contains W .

7. Consider two specific elements of G = GL(n,2)

$$A = \begin{pmatrix} 1 & 0 \\ 1 & 1 \end{pmatrix} \qquad B = \begin{pmatrix} 1 & 1 \\ 0 & 1 \end{pmatrix} .$$

Let H be the subgroup of G generated by A and K be the sub-group of G generated by B . Prove that $H = \{\ldots, A^{-2}, A^{-1}, I, A, A^2, \ldots\}$, and similarly for K .

8. Continuing with exercise 7, show that the product set

$$HK = \{hk \mid h \in H, k \in K\}$$

is not a subgroup of G . (Show that ABAB is not of the form $A^r B^s$.)

9. We say that a subgroup K of G <u>normalizes</u> a subgroup H of G if for each $k \in K$ we have $kHk^{-1} = H$. Prove that if K normalizes H , then KH is a subgroup of G .

10. We can define an injective map

$$\phi : \mathbb{C} \to M_2(\mathbb{R})$$

as follows: represent $\alpha \in \mathbb{C}$ as $\alpha = \rho e^{i\theta}$ with $\rho \geq 0$ and set

$$\phi(\alpha) = \sqrt{\rho} \begin{pmatrix} \cos\theta & \sin\theta \\ -\sin\theta & \cos\theta \end{pmatrix} .$$

Show that $\phi(\alpha\beta) = \phi(\alpha)\phi(\beta)$ but that $\phi(\alpha+\beta)$ need not equal $\phi(\alpha)+\phi(\beta)$.

11. Let G be the multiplicative group of complex numbers of unit length. We say that $\alpha \in G$ is a <u>primitive</u> n^{th} <u>root of unity</u> if $\alpha^n = 1$, but none of $\alpha, \alpha^2, \ldots, \alpha^{n-1}$ are equal to one. Show that an isomorphism of G onto itself must send primitive n^{th} roots of unity to primitive n^{th} roots of unity for each n . For each n, how many primitive n^{th} roots of unity are there in G ?

12. Let $\alpha = (a_1, a_2, a_3)$ and $\beta = (b_1, b_2, b_3)$ be two elements in R^3 . Take the two "purely imaginary" quaternions

$$\alpha' = a_1 i + a_2 j + a_3 k$$

$$\beta' = b_1 i + b_2 j + b_3 k \quad .$$

Show that if α' and β' are multiplied as quaternions, then

$$\alpha'\beta' - \text{real part } (\alpha'\beta')$$

is just the usual cross product of vectors in R^3 .

Chapter 2
Orthogonal Groups

A. Inner products

We have a consistent notion of conjugation for $R \subset C \subset H$. Namely,

$$\text{for } x \in R , \quad \bar{x} = x .$$

$$\text{For } \alpha = x + iy \in C , \quad \bar{\alpha} = x - iy .$$

$$\text{For } q = x + iy + jz + kw \in H , \quad \bar{q} = x - iy - jz - kw .$$

We clearly have $\bar{\bar{\alpha}} = \alpha$ in all cases and

$$\overline{(\alpha + \beta)} = \bar{\alpha} + \bar{\beta} .$$

It is an exercise to prove that

$$\overline{\alpha\beta} = \bar{\beta}\bar{\alpha} .$$

Of course for R or C this is the same as

$$\overline{\alpha\beta} = \bar{\alpha}\bar{\beta} .$$

Let $k \in \{R, C, H\}$ and define an __inner product__ $\langle \ , \ \rangle$ on k^n by

$$\langle x,y \rangle = x_1 \bar{y}_1 + \ldots + x_n \bar{y}_n \quad .$$

Proposition 1: $\langle \ , \ \rangle$ has the following properties:

(i) $\langle x, y+z \rangle = \langle x,y \rangle + \langle x,z \rangle$

(ii) $\langle x+y, z \rangle = \langle x,z \rangle + \langle y,z \rangle$

(iii) $a\langle x,y \rangle = \langle ax,y \rangle \ , \quad \langle x, ay \rangle = \langle x,y \rangle \bar{a}$

(iv) $\overline{\langle x,y \rangle} = \langle y,x \rangle$

(v) $\langle x,x \rangle$ is always a real number ≥ 0 and
$\langle x,x \rangle = 0 \Leftrightarrow x = (0, \ldots, 0)$.

(vi) If $e_1, \ldots, e_n$ is the standard basis for k^n
$(e_i = (0, \ldots, 0, 1, 0, \ldots, 0))$, then
$$\langle e_i, e_j \rangle = \delta_{ij} = \begin{cases} 1 & \text{if } i = j \\ 0 & \text{if } i \neq j \end{cases} .$$

(vii) The inner product is nondegenerate; i.e.,

If $\langle x,y \rangle = 0$ for all y , then $x = (0, \ldots, 0)$;
If $\langle x,y \rangle = 0$ for all x , then $y = (0, \ldots, 0)$.

Proof: Exercise.

Definition: The length $|x|$ of $x \in k^n$ is
$$|x| = \sqrt{\langle x,x \rangle} .$$

Recall that if $A \in M_n(k)$, its conjugate $\bar{A}$ is obtained by replacing each a_{ij} by $\overline{a_{ij}}$; its transpose ${}^t A$ is obtained by replacing each a_{ij} by a_{ji} . These two operations commute so that the symbol

$$t_{\bar{A}}$$

(the conjugate transpose of A) is unambiguous .

Recall that for $\mathbb{H}^n$ we must operate on the right (since we defined (scalar)(vector) on the left). So we do the same for $\mathbb{R}^n$ and $\mathbb{C}^n$. Thus we use row vectors.

Proposition 2: For any $x,y \in k^n$ and $A \in M_n(k)$ we have

$$\langle xA,y \rangle = \langle x,y\,^t\bar{A} \rangle .$$

Proof: Let $A = (a_{ij})$.

$$xA = (x_1 a_{11} + \ldots + x_n a_{n1}, \ldots, x_1 a_{1n} + \ldots + x_n a_{nn})$$

$$y\,^t\bar{A} = (y_1 \bar{a}_{11} + \ldots + y_n \bar{a}_{1n}, \ldots, y_1 \bar{a}_{n1} + \ldots + y_n \bar{a}_{nn})$$

Thus the left hand side $\langle xA,y \rangle$ equals

$$(x_1 a_{11} + \ldots + x_n a_{n1})\bar{y}_1 + \ldots + (x_1 a_{1n} + \ldots + x_n a_{nn})\bar{y}_n ,$$

and the right hand side $\langle x,y\,^t\bar{A} \rangle$ equals

$$x_1 (a_{11}\bar{y}_1 + \ldots + a_{1n}\bar{y}_n) + \ldots + x_n (a_{n1}\bar{y}_1 + \ldots + a_{nn}\bar{y}_n) .$$

It is easy to see that these contain exactly the same terms.

B. Orthogonal groups

Again let $k \in \{\mathbb{R}, \mathbb{C}, \mathbb{H}\}$.

Definition:

$$\mathcal{O}(n,k) = \{A \in M_n(k) \mid \langle xA, yA \rangle = \langle x,y \rangle \quad \text{for all} \quad x,y \in k^n \} \ .$$

Proposition 3: $\mathcal{O}(n,k)$ is a group.

Proof: If $A, B \in \mathcal{O}(n,k)$, then

$$\langle xAB, yAB \rangle = \langle xA, yA \rangle = \langle x,y \rangle$$

so that

$$AB \in \mathcal{O}(n,k) \ .$$

Clearly the identity matrix I is in $\mathcal{O}(n,k)$.

If $A \in \mathcal{O}(n,k)$ we have

$$\langle e_i A, e_j A \rangle = \langle e_i, e_j \rangle = \delta_{ij} = \begin{cases} 1 & \text{if} \quad i = j \\ 0 & \text{if} \quad i \neq j \end{cases} \ .$$

Now $e_i A$ is just the ith row of A and we see that $\langle e_i A, e_j A \rangle$ is just the ij entry in the product

$$A^t\bar{A} \ .$$

Thus $A^t\bar{A} = I$. But then $^t\bar{A}A$ is also the identity since $^t(\overline{^t\bar{A}A}) = {^t}(\bar{A}\,^tA) = A^t\bar{A} = I$. Thus $^t\bar{A} = A^{-1}$, a left hand and right inverse for A. (More generally, we saw in section C of chapter I that for matrices a left inverse was automatically a right inverse.) Finally,

$$\langle xA^{-1}, yA^{-1} \rangle = \langle xA^{-1}A, yA^{-1}A \rangle = \langle x,y \rangle \ ,$$

showing that $A^{-1} \in \mathcal{O}(n,k)$.   q.e.d.

Definition: For $k = R$ we write $\mathfrak{G}(n,k)$ as $\mathfrak{G}(n)$ and call it the orthogonal group. For $k = C$ we write it as $U(n)$ and call it the unitary group. For $k = H$ we write it as $Sp(n)$ and call it the symplectic group.

Proposition 4: Let $A \in M_n(k)$. Then the following conditons are equivalent:

(i) $A \in \mathfrak{G}(n,k)$

(ii) $\langle e_i A, e_j A \rangle = \delta_{ij}$

(iii) A sends orthonormal bases to orthonormal bases

(iv) The rows of A form an orthonormal basis

(v) The columns of A form an orthonormal basis

(vi) ${}^t\bar{A} = A^{-1}$.

Proof: Exercise.

Proposition 5: Let $A \in M_n(R)$. Then $A \in \mathfrak{G}(n) \Leftrightarrow A$ preserves lengths.

Proof: A preserves lengths $\Leftrightarrow \langle xA, xA \rangle = \langle x,x \rangle$ for all $x \in R^n$. So $\Rightarrow$ is trivial. Conversely, we have

$$\langle (x+y)A, (x+y)A \rangle = \langle x+y, x+y \rangle = \langle x,x \rangle + \langle x,y \rangle + \langle y,x \rangle + \langle y,y \rangle$$

$$= \langle xA, xA \rangle + \langle xA, yA \rangle + \langle yA, xA \rangle + \langle yA, yA \rangle .$$

This gives $\langle x,y \rangle + \langle y,x \rangle = \langle xA, yA \rangle + \langle yA, xA \rangle$ and since $\langle \ , \ \rangle$ over R is symmetric, this proves

$$\langle xA, yA \rangle = \langle x, y \rangle \ , \ \text{i.e.,} \quad A \in \mathfrak{G}(n) \ .$$

Proposition 5 **bis:** **Proposition** 5 **also holds for** $\mathbb{C}$ **and** $\mathbb{H}$.

Proof: Calculate $\langle (e_i + e_j)A, (e_i + e_j)A \rangle$ just as above to get

$$\langle e_i A, e_j A \rangle + \langle e_j A, e_i A \rangle = 0 \ .$$

Then consider $x = x_i e_i + x_j e_j$ and calculate $\langle xA, xA \rangle$. We get

$$x_i \overline{x_j} \langle e_i A, e_j A \rangle + x_j \overline{x_i} \langle e_j A, e_i A \rangle = 0 \quad \text{and thus}$$

$$\langle e_i A, e_j A \rangle (x_i \overline{x_j} - \overline{x}_j x_i) = 0$$

and this forces $\langle e_i A, e_j A \rangle = 0$. $\qquad$ q.e.d.

Let us look at $\mathfrak{G}(n)$, $U(n)$ and $Sp(n)$ for small n . $\mathfrak{G}(1)$ is the set of all real numbers of length one, so $\mathfrak{G}(1) = \{1, -1\}$. $U(1)$ is just the set of all complex numbers of length one. This is the circle group S^1 . $Sp(1)$ is the group of all quaternions of unit length. If we define

$$S^{k-1} = \{x \in \mathbb{R}^k | \ |x| = 1\}$$

to be the unit $(k-1)$-sphere we see that

$$\mathfrak{G}(1) = S^0 \ , \quad U(1) = S^1 \ , \quad Sp(1) = S^3 \ .$$

It is an interesting fact that these are the only spheres which can be groups.

Proposition 6: If $k \in \{\mathbb{R}, \mathbb{C}\}$ and $A \in \mathfrak{G}(n,k)$, then

$$(\det A)(\overline{\det A}) = 1 \ .$$

Proof: $A\,{}^{t}\bar{A} = I \Rightarrow (\det A)(\det\,{}^{t}\bar{A}) = 1$, and clearly

$$\det\,{}^{t}\bar{A} = \det\,\bar{A} = \overline{\det A} \ . \qquad\qquad \text{q.e.d.}$$

Thus if $A \in \mathfrak{O}(n)$ $(= \mathfrak{O}(n,\mathbb{R}))$, then $\det A \in \{1,-1\}$. We define

$$SO(n) = \{A \in \mathfrak{O}(n) \mid \det A = 1\}$$

and call this the special orthogonal group (also called the rotation group). Similarly, we define

$$SU(n) = \{A \in U(n) \mid \det A = 1\}$$

and call this the special unitary group.

An example of an element of $\mathfrak{O}(2) - SO(2)$ is $\begin{pmatrix} 1 & 0 \\ 0 & -1 \end{pmatrix}$. This sends $e_1 = (1,0)$ to e_1 and sends $e_2 = (0,1)$ to $-e_2$. It is just the reflection in the first axis, and has determinant equal to -1 .

C. The isomorphism question

At the end of Chapter I we showed that two groups which were defined quite differently were isomorphic. We have now defined several series of groups ($GL(n,k)$ for $n = 1,2,\ldots$ and $k \in \{\mathbb{R},\mathbb{C},\mathbb{H}\}$ and $\mathfrak{O}(n,k)$ for $n = 1,2,\ldots$ and $k \in \{\mathbb{R},\mathbb{C},\mathbb{H}\}$) and our major goal is to find out which of these are isomorphic. The basic idea will be to develop invariants of matrix groups (dimension, rank, etc.), i.e., two groups which are isomorphic must have the same invariants. This will make it possible to say that certain groups are not isomorphic. But when two differently defined groups are indeed isomorphic, an isomorphism may be hard to find. This is why we will work so hard to develop invariants--to reduce as much as possible the cases where we must look for isomorphisms. In this section we will give one iso-morphism.

Suppose you suspect that $Sp(1)$ and $SU(2)$ are isomorphic. How would you try to find an isomorphism? $Sp(1)$ is the set of all quaternions of unit length and $SU(2)$ is the set of all complex 2×2 matrices A such that $A^t \bar{A} = I$ and $\det A = 1$. The operatio in $Sp(1)$ is multiplication of quaternions, in $SU(2)$ it is matrix multiplication.

Proposition 7: The map $\phi : M_n(\mathbb{H}) \to M_{2n}(\mathbb{C})$ defined in §D of Chapter I induces an isomorphism

$$\phi : Sp(1) \to SU(2) .$$

Proof: We have seen that ϕ induces an injective homomorphism of $GL(n, \mathbb{H})$ into $GL(2n, \mathbb{C})$, so restriction of ϕ to $Sp(1)$ is still an injective homomorphism. So we just need to show that

(i) $A \in Sp(1) \Rightarrow \phi(A) \in SU(2)$ and

(ii) every $B \in SU(2)$ is some $\phi(A)$ with $A \in Sp(1)$.

If $A = a + ib + jc + kd$ then $\phi(A) = \begin{pmatrix} a+ib & -c-id \\ c-id & a+ib \end{pmatrix}$ so that

$$\phi(A)^t \overline{\phi(A)} = \begin{pmatrix} a+ib & -c-id \\ c-id & a-ib \end{pmatrix} \begin{pmatrix} a-ib & -c+id \\ c+id & a+ib \end{pmatrix} = \begin{pmatrix} 1 & 0 \\ 0 & 1 \end{pmatrix}$$

since $a^2 + b^2 + c^2 + d^2 = 1$. Also $\det \phi(A) = 1$ so $\phi(A) \in SU(2)$.

Let $B = \begin{pmatrix} \alpha & \beta \\ \gamma & \delta \end{pmatrix} \in SU(2)$. Using $\det B = 1$ and the fact that the rows are orthogonal unit vectors, we find that

$$\delta = \bar{\alpha} \quad \text{and} \quad \gamma = \bar{\beta} \; .$$

So, if $\alpha = a + ib$ and $\beta = c - id$, we may take $A = a + ib + jc + kd$ and have $\phi(A) = B$ (and $a^2 + b^2 + c^2 + d^2 = 1$).

D. Reflections in R^n

Let u be a unit vector in R^n and let

$$u^{\perp} = \{x \in R^n \mid \langle x, u \rangle = 0\}$$

be its orthogonal complement.

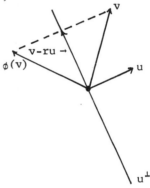

The projection of a vector v into $u^{\perp}$ is to be $v - ru$, where $r \in R$ is to be chosen so that $v - ru$ is in $u^{\perp}$. So $0 = \langle v - ru, u \rangle = \langle v, u \rangle - r\langle u, u \rangle$ and thus

$$r = \langle v, u \rangle \; .$$

Clearly then the reflection of v in $u^{\perp}$ is to be

$$\phi(v) = v - 2ru = v - 2\langle v, u \rangle u \; .$$

Choose an orthonormal basis $u_1, \ldots, u_n$ with $u_1 = u$. Then, using this basis, the reflection ϕ is given by the matrix

$$\begin{pmatrix} -1_1 & & \text{O} \\ \text{O} & 1 & \\ & & \ddots \\ & & & 1 \end{pmatrix}.$$

Let A be the linear map of R^n given by sending $e_1, \ldots, e_n$ to $u_1, \ldots, u_n$. By Proposition 4 A is orthogonal. So relative to our standard basis $e_1, \ldots, e_n$ the reflection ϕ is given by

$$A \begin{pmatrix} -1_1 & & \bigcirc \\ & \ddots & \\ \bigcirc & & 1 \end{pmatrix} A^{-1} = A \begin{pmatrix} -1_1 & & \bigcirc \\ & \ddots & \\ \bigcirc & & 1 \end{pmatrix} {}^t A \quad .$$

Conversely we see that such a matrix represents a reflection in the orthogonal complement of the vector $e_1 A$.

In $\mathbb{R}^2$ let the unit vector u be written as

$$u = (\cos \alpha, \sin \alpha) \quad .$$

Then $(-\sin \alpha, \cos \alpha)$ is a unit vector in $u^{\perp}$. The matrix A sending e_1 to u and e_2 to $(-\sin \alpha, \cos \alpha)$ must satisfy

$$(1,0) \begin{pmatrix} a_{11} & a_{12} \\ a_{21} & a_{22} \end{pmatrix} = (\cos \alpha, \sin \alpha)$$

$$(0,1) \begin{pmatrix} a_{11} & a_{12} \\ a_{21} & a_{22} \end{pmatrix} = (-\sin \alpha, \cos \alpha)$$

so

$$A = \begin{pmatrix} \cos \alpha & \sin \alpha \\ -\sin \alpha & \cos \alpha \end{pmatrix} \quad .$$

Thus the matrix giving reflection in $u^{\perp}$ is

$$\Phi = \begin{pmatrix} \cos \alpha & \sin \alpha \\ -\sin \alpha & \cos \alpha \end{pmatrix} \begin{pmatrix} -1 & 0 \\ 0 & 1 \end{pmatrix} \begin{pmatrix} \cos \alpha & -\sin \alpha \\ \sin \alpha & \cos \alpha \end{pmatrix}$$

$$\Phi = \begin{pmatrix} -\cos 2\alpha & \sin 2\alpha \\ \sin 2\alpha & \cos 2\alpha \end{pmatrix} \quad .$$

The matrix A is easily seen to be a rotation of R^2 through an angle α .

Later in this course we will prove that $\mathbb{O}(n)$ is generated by reflections--that is, any element of $\mathbb{O}(n)$ may be obtained by a finite sequence of reflections.

E. Exercises

1. Prove Proposition 1.

2. Prove Proposition 4.

3. Let A be any element of $\mathbb{O}(n)$ with det A = -1 . Show that

$$\mathbb{O}(n) - SO(n) = \{BA \mid B \in SO(n)\} .$$

4. Show that any element of SO(2) can be written as

$$\begin{pmatrix} \cos \theta & \sin \theta \\ -\sin \theta & \cos \theta \end{pmatrix} .$$

5. If $A \in U(n)$ and $\lambda \in C$ has length one, show that $\lambda A \in U(n)$.

6. Let L_1 and L_2 be lines through the origin in R^2 . Show that reflection in L_1 followed by reflection in L_2 equals a rotation through twice the angle between L_1 and L_2 .

7. A matrix $A \in M_n(R)$ is said to be <u>idempotent</u> if AA = A . Show that the image of R^n under A is precisely the fixed-point set of A . Such a map is called a <u>projection</u> of R^n onto its image.

What can you say about the determinant of a idempotent matrix ? What is the image of R^2 under $A = \begin{pmatrix} \frac{1}{2} & \frac{1}{2} \\ \frac{1}{2} & \frac{1}{2} \end{pmatrix}$.

8. A matrix A is <u>nilpotent</u> if some power of it is the zero matrix. For example $A = \begin{pmatrix} 0 & a & b \\ 0 & 0 & c \\ 0 & 0 & 0 \end{pmatrix}$ has its third power zero. Prove that a nilpotent matrix is singular. Prove that any $A = (a_{ij})$ with

$$a_{ij} = 0 \quad \text{whenever} \quad i > j$$

is nilpotent. Find two nilpotent matrices A and B whose product AB is not nilpotent.

9. Let U be the set of all matrices $A = (a_{ij})$ with all diagonal elements equal to one and

$$a_{ij} = 0 \quad \text{whenever} \quad i > j \ .$$

Prove that U is a group under matrix multiplication (the group of <u>unipotent</u> matrices in $M_n(R)$) .

Chapter 3
Homomorphisms

A. Curves in a vector space

We are going to define our first invariant of a matrix group, its
dimension. Matrix groups whose dimensions are different can't be
isomorphic. The dimension of a matrix group is going to be the dimension of its space of tangent vectors (a vector space), so we first
define these.

Let V be a finite-dimensional real vector space. By a curve
γ in V we mean a continuous function $γ:(a,b) → V$ where (a,b)
is an open interval in $\mathbb{R}$.

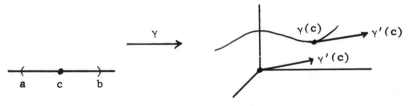

For $c \in (a,b)$ we say γ is differentiable at c if

$$\lim_{h \to 0} \frac{γ(c+h) - γ(c)}{h}$$

exists. When this limit exists, it is a vector in V . We denote it
by $γ'(c)$ and call it the tangent vector to γ at $γ(c)$.

It is a standard result from calculus that if we choose a basis for V and thus represent γ as $(\gamma_1, \ldots, \gamma_n)$ (γ_i's being real valued), then $\gamma'(c)$ exists if and only if each $\gamma_i'(c)$ exists and

$$\gamma'(c) = (\gamma_1'(c), \ldots, \gamma_n'(c)) \ .$$

Now $M_n(\mathbb{R})$, $M_n(\mathbb{C})$, $M_n(\mathbb{H})$ can all be considered to be real vector spaces (of dimensions n^2, $2n^2$ and $4n^2$). If G is a matrix group in $M_n(k)$ then a <u>curve in</u> G is a curve in $M_n(k)$ with all values $\gamma(u)$ for $u \in (a,b)$ lying in G .

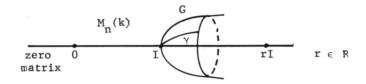

Suppose we have curves $\gamma, \sigma : (a,b) \to G$. Then we can define a new curve, the <u>product curve</u>, by

$$(\gamma\sigma)(u) = \gamma(u)\sigma(u) \ .$$

<u>Proposition 1</u>: <u>Let</u> $\gamma, \sigma : (a,b) \to G$ <u>be curves, both of which are</u> <u>differentiable at</u> $c \in (a,b)$. <u>Then the product curve</u> $\gamma\sigma$ <u>is dif-</u> <u>ferentiable at</u> c <u>and</u>

$$(\gamma\sigma)'(c) = \gamma(c)\sigma'(c) + \gamma'(c)\sigma(c) \ .$$

<u>Proof</u>: Let $\gamma(u) = (\gamma_{ij}(u))$, $\sigma(u) = (\sigma_{ij}(u))$. Then

$$(\gamma\sigma)(u) = (\sum_k \gamma_{ik}(u)\sigma_{kj}(u)) \ ,$$

so that

$$(\gamma\sigma)'(u) = (\sum_k \{v'_{ik}(u)\sigma_{kj}(u) + v_{ik}(u)\sigma'_{kj}(n)\})$$

$$= \gamma'(u)\sigma(u) + \gamma(u)\sigma'(u) .$$

Proposition 2: Let G be a matrix group in $M_n(k)$. Let T be the set of all tangent vectors $\gamma'(0)$ to curves $\gamma : (a,b) \to G$, $\gamma(0) = I$ $(0 \in (a,b))$. Then T is a subspace of $M_n(k)$.

Proof: If $\gamma'(0)$ and $\sigma'(0)$ are in T, then

$$(\gamma\sigma)(0) = \gamma(0)\sigma(0) = II = I \quad \text{and}$$

$$(\gamma\sigma)'(0) = \gamma'(0)\sigma(0) + \gamma(0)\sigma'(0) = \gamma'(0) + \sigma'(0) .$$

Thus T is closed under vector addition.

T is also closed under scalar multiplication, for if $\gamma'(0) \in T$ and $r \in R$, let

$$\sigma(u) = \gamma(ru) .$$

Then $\sigma(0) = \gamma(0) = I$, σ is differentiable and $\sigma'(0) = r\gamma'(0)$. Since $M_n(k)$ is a finite dimensional vector space, so is T.

Definition: If G is a matrix group, its <u>dimension</u> is the dimension of the vector space T (of tangent vectors to G at I).

Example 1: $U(1)$ has dimension 1.

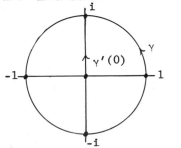

Example 2: dim Sp(1) = 3 .

Let $\gamma : (a,b) \to Sp(1)$ be a smooth curve with $\gamma(0) = 1$. Then $\gamma'(0)$ will be an element of $\mathbb{H} = R^4$. We first show $\gamma'(0)$ is in the span of i,j,k ; i.e. it is a quaternion with zero real part. Let

$$\gamma(t) = x(t) + i\ y(t) + j\ z(t) + k\ w(t)$$

with $x(0) = 1$ and $y(0) = 0$, $z(0) = 0$, $w(0) = 0$. We note that $x(0)$ is a maximum for the function x so that $\gamma'(0)$ $= 0 + i\ y'(0) + j\ z'(0) + k\ w'(0)$, as asserted.

Conversely, let $q = i\mu + j\nu + k\lambda$ be any quaternion with zero real part. We claim that there exists a smooth curve γ in Sp(1) such that $\gamma'(0) = q$. Indeed,

$$\gamma(t) = \sqrt{1 - \sin^2 \mu t - \sin^2 \nu t - \sin^2 \lambda t} + i\ \sin \mu t + j\ \sin \nu t + k\ \sin kt$$

can be readily verified to be such a curve (which is defined on some interval $[0,\varepsilon)$, i.e. for t small).

Example 3: dim $GL(n, R) = n^2$.

The determinant function $\det : M_n(R) \to R$ is continuous and $\det(I) = 1$. So there is some ε-ball about I in $M_n(R)$ such that for each A in this ball $\det A \neq 0$; i.e., $A \in GL(n, R)$. If v is any vector in $M_n(R)$ define a curve σ in $M_n(R)$ by

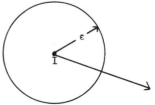

$$\sigma(t) = tv + I .$$

Then $\sigma(0) = I$ and $\sigma'(0) = v$ and for small t, $\sigma(t)$ is in $GL(n,R)$. Hence the tangent space T is all of $M_n(R)$ which has dimension n^2.

A similar argument shows that $\dim GL(n,C) = 2n^2$.

We will now get upper bounds for the dimensions of $O(n)$, $U(n)$ and $Sp(n)$ after a few preliminaries.

Definition: $A \in M_n(R)$ is said to be skew-symmetric if

$$A + {}^tA = 0 ;$$

i.e. if $a_{ij} = -a_{ji}$ for each i,j . In particular, the diagonal terms must all be zero.

Proposition 3: Let $so(n)$ denote the set of all skew-symmetric matrices in $M_n(R)$. Then $so(n)$ is a linear subspace of $M_n(R)$, and its dimension is $\frac{n(n-1)}{2}$.

Proof: The zero matrix is in $so(n)$, and if A,B belong to $so(n)$, then

$$(A + B) + {}^t(A + B) = A + {}^tA + B + {}^tB = 0 ,$$

so that $so(n)$ is closed under vector addition. It is also closed under scalar multiplication, for is $A \in so(n)$ and $r \in R$, then ${}^t(rA) = r{}^tA$ so that $rA + {}^t(rA) = r(A + {}^tA) = 0$.

To check the dimension of $so(n)$ we get a basis. Let E_{ij} denote the matrix whose entries are all zero except the ij entry, which is 1 , and the ji entry, which is -1 . If we define these

E_{ij} only for $i < j$, it is easy to see that they form a basis for so(n) , and it is easy to count that there are

$$(n-1) + (n-2) +\ldots+ 1 = \frac{n(n-1)}{2} \qquad \text{of them .}$$

Definition: A matrix $B \in M_n(\mathbb{C})$ is skew-Hermitian if

$$B + {}^t\bar{B} = 0 .$$

Thus if $b_{jk} = c + id$, then $\bar{b}_{kj} = -b_{jk} = -c - id$ and $b_{kj} = -c + id$. In particular if $j = k$ we have $c + id = -c + id$, so that the diagonal terms of a skew-Hermitian matrix are purely imaginary.

Let su(n) be the set of skew-Hermitian matrices in $M_n(\mathbb{C})$. By the observation just made we see that su(n) is not a vector space over $\mathbb{C}$.

Proposition 4: $su(n) \subset M_n(\mathbb{C})$ is a real vector space of dimension

$$n + 2 \frac{n(n-1)}{2} = n^2 .$$

Proof: Exercise.

We make a similar definition for matrices in $M_n(\mathbb{H})$, and call $C \in M_n(\mathbb{H})$ skew-symplectic if

$$C + {}^t\bar{C} = 0 .$$

In the exercises one shows that the set sp(n) of such matrices is a real vector space of dimension

$$3n + 4 \frac{n(n-1)}{2} = n(2n+1) .$$

Proposition 5: If β is a curve through the identity $(\beta(0) = I)$

in O(n)   then $\beta'(0)$ is skew-symmetric

in U(n)   then $\beta'(0)$ is skew-Hermitian

in Sp(n)   then $\beta'(0)$ is skew-symplectic .

Proof: In each case we have that the product curve is constant

$$\beta(u)^{t}\bar{\beta}(u) = I .$$

Thus its derivative is zero, and the result follows from Proposition 1.

Corollary:

$$\text{Dim } O(n) \leq \frac{n(n-1)}{2}$$

$$\text{Dim } U(n) \leq n^2$$

$$\text{Dim } Sp(n) \leq n(2n+1) .$$

Later we will show that these are equalities.

B.   Smooth homomorphisms

Let $\phi : G \to H$ be a homomorphism of matrix groups. Since G and
H are in vector spaces, it is clear what it means for ϕ to be continuous. From now on homomorphism always means continuous homomorphism. This being so, a curve

$$\rho : (a,b) \to G$$

gives a curve $\phi \circ \rho : (a,b) \to H$ by $(\phi \circ \rho)(u) = \phi(\rho(u))$ in H .

Definition: A homomorphism $\phi : G \to H$ of matrix groups is smooth if for every differentiable curve ρ in G , $\phi \circ \rho$ is differentiable.

Definition: Let $\phi : G \to H$ be a smooth homomorphism of matrix groups. If $\gamma'(0)$ is a tangent vector to G at I we define a tangent vector $d\phi(\gamma'(0))$ to H at I by

$$d\phi(\gamma'(0)) = (\phi \circ \rho)'(0) .$$

The resulting map $d\phi : T_G \to T_H$ is called the differential of ϕ .

Proposition 6: $d\phi : T_G \to T_H$ is a linear map .

Proof: If $\rho'(0)$ and $\sigma'(0)$ are in T_G , consider

$$d\phi(a\rho'(0) + b\sigma'(0))$$

with $a,b \in \mathbb{R}$. By definition this equals

$$\{ \phi \circ (a\rho + b\sigma) \}'(0) = \{ a(\phi \circ \rho) + b(\phi \circ \sigma) \}'(0)$$

$$= a(\phi \circ \rho)'(0) + b(\phi \circ \sigma)'(0) = a\ d\phi(\rho'(0)) + b\ d\phi(\sigma'(0)) ,$$

proving that $d\phi$ is linear.

Proposition 7: If $G \overset{\phi}{\to} H \overset{\psi}{\to} K$ are smooth homomorphisms, then so is $\psi \circ \phi$ and

$$d(\psi \circ \phi) = d\psi \circ d\phi .$$

Proof: The first part is obvious. For the second, let $\gamma'(0)$ be a tangent vector of G . Then

$$d(\psi \circ \phi)(\gamma'(0)) = (\psi \circ \phi \circ \gamma)'(0) = d\psi(\phi \circ \gamma)'(0) = d\psi \circ d\phi(\gamma'(0)) .$$

Corollary: If $\phi : G \to H$ is a smooth isomorphism, then

$d\phi : T_G \to T_H$ <u>is a linear isomorphism</u> <u>and</u> $\dim G = \dim H$.

<u>Proof</u>: $\phi^{-1} \circ \phi$ is the identity, so $d\phi^{-1} \, d\phi : T_G \to T_G$ is the identity. Thus $d\phi$ is injective and $d\phi^{-1}$ is surjective. $\phi \, \phi^{-1}$ is the identity, so $d\phi \, d\phi^{-1} : T_H \to T_H$ is the identity. Thus $d\phi^{-1}$ is injective and $d\phi$ is surjective. q.e.d.

C. Exercises

1. Let $\gamma : (-1,1) \to M_3(\mathbb{R})$ be given by

$$\gamma(t) = \begin{pmatrix} \cos t & \sin t & 0 \\ -\sin t & \cos t & 0 \\ 0 & 0 & 1 \end{pmatrix} .$$

Show that γ is a curve in $SO(3)$ and find $\gamma'(0)$. Show that $(\gamma^2)'(0) = 2\gamma'(0)$.

2. Let $\sigma : (-1,1) \to M_3(\mathbb{R})$ be given by

$$\sigma(t) = \begin{pmatrix} 0 & 0 & 0 \\ 0 & \cos t & \sin t \\ 0 & -\sin t & \cos t \end{pmatrix} .$$

Calculate $\sigma'(0)$. Write the matrix for $\gamma(t)\sigma(t)$ and verify that $(\gamma\sigma)'(0) = \gamma'(0) + \sigma'(0)$.

3. Let $\rho : (-1,1) \to M_3(\mathbb{C})$ be given by

$$\rho(t) = \begin{pmatrix} e^{i\pi t} & 0 & 0 \\ 0 & e^{i\frac{\pi t}{2}} & 0 \\ 0 & 0 & e^{i\frac{\pi t}{2}} \end{pmatrix}.$$

Show that ρ is a curve in $U(3)$. Calculate $\rho'(0)$.

4. Let $\alpha : (-1,1) \to \mathbb{H}$ be defined by

$$\alpha(t) = (\cos t)j + (\sin t)k .$$

Show that α is in $Sp(1)$ and calculate $\alpha'(t)$.

5. Let H be a subgroup of a matrix group G . Show that T_H is a linear subspace of T_G so that dim H ≤ dim G .

6. Show that the set sp(n) of n × n skew-symplectic matrices is a real vector space and calculate its dimension.

7. Let T be the set of upper triangular matrices in $M_n(\mathbb{R})$. That is, $A = (a_{ij}) \in T$ if and only if $a_{ij} = 0$ whenever i > j . Show that T is a linear subspace of $M_n(\mathbb{R})$ and calculate its dimension. Show that T is a subalgebra of $M_n(\mathbb{R})$ (i.e. show that T is closed under multiplication). Show that $A \in T$ is nonsingular (i.e. is a unit) if and only if each $a_{ii} \neq 0$. (Note that the group U defined in Exercise 9 of Chapter 2 is a subgroup of the group of units in the algebra T .)

Chapter 4
Exponential and Logarithm

A. Exponential of a matrix

Given a matrix group G we have defined a vector space T -- the
tangent space to G at I . In this chapter we develop maps to send
T to G and G to T and study their properties. We will work
with real matrices -- developments for C and H are quite analogous.
(We need these maps to determine dimensions of some of our matrix
groups.)

Definition: Let A be a real n × n matrix and set

$$e^A = I + A + \frac{A^2}{2!} + \frac{A^3}{3!} + \dots$$

where A^2 means the matrix product AA , etc. We say that this
sequence converges if each of the n^2 real-number sequences

$$(I)_{ij} + (A)_{ij} + (\frac{A^2}{2!})_{ij} + (\frac{A^3}{3!})_{ij} + \dots$$

converges.

Proposition 1: For any real n × n matrix A, the sequence

$$I + A + \frac{A^2}{2!} + \dots$$

converges.

Proof: Let m be the largest $|a_{ij}|$ in A. Then:

The biggest element in the first term is 1.

The biggest element in the second term is m.

The biggest element in the third term is $\leq \dfrac{nm^2}{2!}$.

The biggest element in the fourth term is $\leq \dfrac{n^2 m^3}{3!}$, etc.

Any ij sequence is dominated by 1, m, $\dfrac{nm^2}{2!}$, $\dfrac{n^2 m^3}{3!}$, ...,

$\dfrac{n^{k-2} m^{k-1}}{(k-1)!}$, ...

Applying the ratio test to this maximal sequence gives

$$\frac{n^{k-1} m^k}{k!} \frac{(k-1)!}{n^{k-2} m^{k-1}} = \frac{nm}{k} .$$

Since n and m are fixed, the ratio goes to 0 as $k \to \infty$, proving (absolute) convergence.

This exponential behaves somewhat like the familiar e^x $(x \in R)$. For if 0 is the zero matrix, we have

$$e^0 = I .$$

Also:

Proposition 2: If the matrices A and B commute, then

$$e^{A+B} = e^A e^B .$$

Proof: We will just indicate a proof by looking at the first few terms.

$$e^{A+B} = I + A + B + \frac{A^2}{2} + AB + \frac{B^2}{2} + \frac{A^3}{6} + \frac{A^2B}{2} + \frac{AB^2}{2} + \frac{B^3}{6} + \dots$$

$$e^A e^B = (I + A + \frac{A^2}{2} + \frac{A^3}{6} + \dots)(I + B + \frac{B^2}{2} + \frac{B^3}{6} + \dots)$$

$$= I + A + B + \frac{A^2}{2} + AB + \frac{B^2}{2} + \frac{A^3}{6} + \frac{A^2B}{2} + \frac{AB^2}{2} + \frac{B^3}{6} + \dots \ .$$

Corollary: For any real $n \times n$ matrix A, e^A is nonsingular.

Proof: A and $-A$ commute, so $I = e^0 = e^{A-A} = e^A e^{-A}$ and thus $1 = (\det e^A)(\det e^{-A})$ and $\det e^A \neq 0$.

From this corollary we see that the map $\exp : M_n(\mathbb{R}) \to M_n(\mathbb{R})$, $\exp(A) = e^A$, actually maps $M_n(\mathbb{R})$ into $GL(n, \mathbb{R})$.

Proposition 3: If A is a real skew-symmetric matrix, then e^A is orthogonal.

Proof: We have $I = e^0 = e^{A + {}^tA} = e^A e^{{}^tA} = (e^A){}^t(e^A)$, proving that e^A is orthogonal.

So, if $so(n) \subset M_n(\mathbb{R})$ is the subspace of skew-symmetric matrices, we see that

$$\exp : so(n) \to \mathbb{O}(n) \ .$$

It is important to note two things which Proposition 3 does not say: (i) it does not say that every orthogonal matrix is some e^A with A skew-symmetric (i.e. it does not say $\exp : so(n) \to \mathbb{O}(n)$ is surjective), and (ii) it does not say that e^A orthogonal implies A is skew-symmetric. It is instructive to examine the case $n = 2$ in some detail.

The general 2×2 real skew-symmetric matrix is of the form

$$\alpha = \begin{pmatrix} 0 & x \\ -x & 0 \end{pmatrix}, \quad x \in \mathbb{R}.$$

To calculate e^{α}, we calculate the powers of α. $\alpha^2 = \begin{pmatrix} -x^2 & 0 \\ 0 & -x^2 \end{pmatrix}$,

$\alpha^3 = \begin{pmatrix} 0 & -x^3 \\ x^3 & 0 \end{pmatrix}$, $\alpha^4 = \begin{pmatrix} x^4 & 0 \\ 0 & x^4 \end{pmatrix}$, $\alpha^5 = \begin{pmatrix} 0 & x^5 \\ -x^5 & 0 \end{pmatrix}$, etc. Then

$$e^{\alpha} = \begin{pmatrix} 1 & 0 \\ 0 & 1 \end{pmatrix} + \begin{pmatrix} 0 & x \\ -x & 0 \end{pmatrix} + \frac{1}{2!}\begin{pmatrix} -x^2 & 0 \\ 0 & -x^2 \end{pmatrix} + \frac{1}{3!}\begin{pmatrix} 0 & -x^3 \\ x^3 & 0 \end{pmatrix}$$
$$+ \frac{1}{4!}\begin{pmatrix} x^4 & 0 \\ 0 & x^4 \end{pmatrix} + \frac{1}{5!}\begin{pmatrix} 0 & x^5 \\ -x^5 & 0 \end{pmatrix} + \dots .$$

From the $1,1$ position we get

$$1 - \frac{x^2}{2!} + \frac{x^4}{4!} - \frac{x^6}{6!} + \dots = \cos x, \text{ etc.}$$

We find that

$$e^{\alpha} = \begin{pmatrix} \cos x & \sin x \\ -\sin x & \cos x \end{pmatrix}$$

which is a plane rotation of x radians. Thus for any real 2×2 skew-symmetric matrix α we have

$$\det e^{\alpha} = 1, \text{ i.e., } e^{\alpha} \in S\mathcal{O}(2).$$

Thus, for example, the reflection $\begin{pmatrix} 1 & 0 \\ 0 & -1 \end{pmatrix} \in \mathcal{O}(2)$ could never be obtained this way.

Note also that $e^{\alpha} = I$ does not imply that α is the zero matrix $(\alpha = \begin{pmatrix} 0 & 2\pi \\ -2\pi & 0 \end{pmatrix}$ has $e^{\alpha} = I)$.

We will see later that these results also hold for larger n.

We conclude this section with a simple observation which is sometimes quite useful in computations.

Proposition 4: If A, B are $n \times n$ matrices over $k \in \{\mathbb{R}, \mathbb{C}, \mathbb{H}\}$ and B is nonsingular, then

$$e^{BAB^{-1}} = Be^A B^{-1} .$$

Proof: $(BAB^{-1})^n = (BAB^{-1})(BAB^{-1})\ldots(BAB^{-1}) = BA^n B^{-1}$, and $B(C+D)B^{-1} = BCB^{-1} + BDB^{-1}$; these and the definition of the exponential of a matrix yield the result.

B. Logarithm

Just as e^x is defined for all $x \in R$ and $\log x$ is defined only for $x > 0$, the logarithm of a matrix will be defined only for matrices near to the identity matrix I .

Let X be a real $n \times n$ matrix and set

$$\log X = (X-I) - \frac{(X-I)^2}{2} + \frac{(X-I)^3}{3} - \frac{(X-I)^4}{4} + \ldots .$$

Proposition 5: For X near I this series converges.

Proof: Let $Y = X - I$ and $Y = (y_{ij})$, and suppose each $|y_{ij}| < \epsilon$. Then

$$|(Y)_{ij}| \leq \epsilon , \quad |(\frac{Y^2}{2})_{ij}| \leq \frac{n\epsilon^2}{2} ,$$

$$|(\frac{Y^3}{3})_{ij}| \leq \frac{n^2\epsilon^3}{3} , \quad |(\frac{Y^k}{k})_{ij}| \leq \frac{n^{k-1}\epsilon^k}{k} .$$

The ratio test gives

$$\frac{n^k \epsilon^{k+1}}{k+1} \cdot \frac{k}{n^{k-1}\epsilon^k} = \frac{k}{k+1} n\epsilon \to n\epsilon .$$

So the series converges for any X such that each entry of $X - I$ is $< \frac{1}{n}$ in magnitude.

Proposition 6: In $M_n(R)$ let U be a neighborhood of I on which log is defined and let V be a neighborhood of O such that exp(V) ⊂ U . Then

   (i) for X ∈ U , $e^{\log X} = X$

   (ii) for A ∈ V , $\log e^A = A$.

Proof: We do (ii) first. $A \in V \Rightarrow e^A \in U \Rightarrow \log e^A$ is defined (i.e., the series converges). $e^A - I = A + \frac{A^2}{2!} + \frac{A^3}{3!} + \ldots$. So

$$\log e^A = (A + \frac{A^2}{2!} + \ldots) - \frac{1}{2}(A + \frac{A^2}{2!} + \ldots)^2 + \frac{1}{3}(A + \frac{A^2}{2!} + \ldots)^3 + \ldots$$

$$= A + [\frac{A^2}{2!} - \frac{A^2}{2}] + [\frac{A^3}{6} - \frac{A^3}{2} + \frac{A^3}{3}] + \ldots = A .$$

(i) is similar. $\log X = (X-I) - \frac{(X-I)^2}{2} + \frac{(X-I)^3}{3} - \ldots$.

$$e^{\log X} = \{I + (X-I) - \frac{(X-I)^2}{2} + \ldots\} + \frac{1}{2!}\{(X-I) - \frac{(X-I)^2}{2} + \ldots\}^2$$

$$+ \frac{1}{3!}\{(X-I) - \frac{(X-I)^2}{2} + \ldots\}^3 + \ldots$$

$$= X - \frac{(X-I)^2}{2} + \frac{(X-I)^2}{2} + \{\frac{(X-I)^3}{3} - \frac{(X-I)^3}{2} + \frac{(X-I)^3}{6}\} + \ldots$$

$$= X .$$

Proposition 7: If X and Y are near I and log X and log Y commute, then

$$\log (XY) = \log X + \log Y .$$

So if X is near I and orthogonal, log X is skew-symmetric.

Proof: $e^{\log XY} = XY = e^{\log X} e^{\log Y} = e^{\log X + \log Y}$, and e is one-to-one near O .

Next X and ^tX commute so that log X and log ^tX commute. If X is orthogonal

$$I = X {}^t X$$

so $0 = \log X {}^t X = \log X + \log {}^t X = \log X + {}^t(\log X)$ showing $\log X$

is skew-symmetric.

We will be able to say more after we have done a little topology.
The picture will be

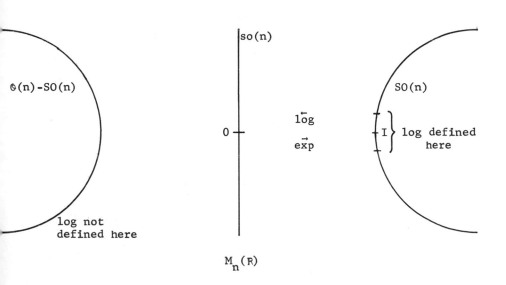

C. <u>One-parameter subgroups</u>

<u>Definition</u>: A <u>one-parameter subgroup</u> γ in a matrix group G
is a smooth homomorphism

$$\gamma : R \to G .$$

Note that it suffices to know γ on some open neighborhood U of 0
in R. For $x \in R$, some $\frac{1}{n} x \in U$ and $\gamma(x) = (\gamma(\frac{1}{n} x))^n$.

Example: Let $k \in \{R, C, H\}$ and $A \in M_n(k)$. Then

$$\gamma(u) = e^{uA} = I + uA + u^2 \frac{A^2}{2!} + \ldots$$

is a one-parameter subgroup of $GL(n,k)$ and $\gamma'(0) = A$.

Proposition 8: Let γ be a one-parameter subgroup of $GL(n,k)$. Then $\exists A \in M_n(k)$ such that

$$\gamma(u) = e^{uA}.$$

Proof: Let $\sigma(u) = \log \gamma(u)$. Then σ is a curve in $M_n(k)$ with

$$\gamma(u) = e^{\sigma(u)}.$$

Let $\sigma'(0) = A$. We just need to show that $\sigma(u)$ is a line through 0 in $M_n(k)$, for then $\sigma(u) = uA$. Hold u fixed.

$$\sigma'(u) = \lim_{v \to 0} \frac{\sigma(u+v) - \sigma(u)}{v} = \lim_{v \to 0} \frac{\log \gamma(u+v) - \log \gamma(u)}{v}$$

$$= \lim_{v \to 0} \frac{\log(\gamma(u)\gamma(v)) - \log \gamma(u)}{v}.$$

Now $u + v = v + u$ and γ is a one-parameter subgroup so that $\gamma(u)$ and $\gamma(v)$ commute. Thus

$$\log(\gamma(u)\gamma(v)) = \log \gamma(u) + \log \gamma(v).$$

So

$$\sigma'(u) = \lim_{v \to 0} \frac{\log \gamma(v)}{v} = \sigma'(0).$$

This proves that $\sigma'(u)$ is independent of u so $\sigma(u)$ is indeed a line through 0 in $M_n(k)$.

So any tangent vector to GL(n,k) is the derivative at 0 of some one-parameter subgroup. We will see now that this is also true for the orthogonal groups $\mathcal{O}(n,k)$.

Proposition 9: Let A be a tangent vector to $\mathcal{O}(n,k)$. Then there exists a unique one-parameter subgroup γ in $\mathcal{O}(n,k)$ such that

$$A = \gamma'(0) .$$

Proof: By definition $A = \rho'(0)$ where ρ is a curve in $\mathcal{O}(n,k)$. Thus

$$\rho(u) \; {}^t\overline{\rho(u)} = I$$

so that

$$\rho'(0) + {}^t\overline{\rho'(0)} = 0 , \text{ i.e.,}$$

$$A + {}^t\bar{A} = 0 .$$

Now $\gamma(u) = e^{uA}$ is a one-parameter subgroup of GL(n,k) , but it lies in $\mathcal{O}(n,k)$ because

$$\gamma(u) \; {}^t\overline{\gamma(u)} = e^{uA} e^{u{}^t\bar{A}} = e^{u(A+{}^t\bar{A})} = I .$$

This proves the proposition. So we have (for GL(n,k) and $\mathcal{O}(n,k)$) a one-to-one correspondence between tangent vectors and one-parameter subgroups.

Taking k = R we have that the tangent space to $\mathcal{O}(n) = \mathcal{O}(n,R)$ is so(n), the vector space of all skew-symmetric $n \times n$ matrices. Thus dim $\mathcal{O}(n)$ = dim so(n) = $\frac{n(n-1)}{2}$.

Taking k = C we have that the tangent space to $U(n) = \mathcal{O}(n,C)$

is su(n), the vector space of all skew-Hermitian n x n complex matrices. Thus

$$\dim U(n) = \dim su(n) = n^2 \ .$$

Taking k = ℍ we get

$$\dim Sp(n) = n(2n+1) \ .$$

What about the dimensions of SO(n) and SU(n)? We will see in Chapter VI (Proposition 3) that the tangent space to SO(n) is again so(n) , so the dimension of SO(n) is also $\frac{n(n-1)}{2}$. But the dimension of SU(n) is one less than the dimension of U(n) . The proof of this must also be deferred to a later chapter, but we will indicate here the result on which it is based.

Definition: The trace of a matrix A = (a_{ij}) is the sum of the diagonal terms;

$$Tr(A) = a_{11} + a_{22} + \ldots + a_{nn} \ .$$

We clearly have

(i) Tr(A+B) = Tr(A) + Tr(B) , and Tr(aA) = a Tr(A) , (so
     Tr is linear) .

Now suppose A = (a_{ij}) is real or complex. Then

(ii) Tr(AB) = Tr(BA) .

To prove (ii) we just write it out. The sum of the diagonal terms in AB is

$$(a_{11}b_{11}+\ldots+a_{1n}b_{n1}) + (a_{21}b_{12}+\ldots+a_{2n}b_{n2}) +\ldots+ (a_{n1}b_{1n}+\ldots+a_{nn}b_{nn})$$

and the sum of the diagonal terms in BA is

$$(b_{11}a_{11}+\ldots+b_{1n}a_{n1}) + (b_{21}a_{12}+\ldots+b_{2n}a_{n2}) +\ldots+ (b_{n1}a_{1n}+\ldots+b_{nn}a_{nn}) \; .$$

Since R and C are commutative one easily checks that these are equal.

Clearly

(iii) $Tr(I) = n$.

Also

(iv) If B is nonsingular, then

$$Tr(BAB^{-1}) = Tr(A) \; .$$

Proof: By (ii) , $Tr(B(AB^{-1})) = Tr((AB^{-1})B) = Tr(AI) = TrA$.

Now we come to the crucial relation.

Theorem: If A is a real or complex matrix, then

$(+)$
$$e^{Tr(A)} = \det (e^{A}) \; .$$

We will prove this later, but a few comments are in order here. First off, $(+)$ looks wrong because the left hand side depends only on the diagonal elements of A and it is not immediately clear that this is true for the right-hand side. The point is that   det and e are also invariant under conjugation just as (iv) for Tr ; so if B is nonsingular

$$\det (e^{BAB^{-1}}) = \det (Be^{A}B^{-1}) = \det (e^{A}) \; .$$

We will prove $(+)$ once we have found how to put matrices in simpler forms by conjugation (in Chapter VIII).

Suppose we know (†). The linear map

$$\text{Tr} : u(n) \to \mathbb{C}$$

actually maps into $i\mathbb{R} \subset \mathbb{C}$ since all diagonal terms in a skew-Hermitian matrix are purely imaginary. It is easy to see that $\text{Tr}(u(n))$ i all of $i\mathbb{R}$. From the rank theorem in linear algebra we know that (all vector spaces being over $\mathbb{R}$ now)

$$\dim u(n) = \dim \text{Tr}(u(n)) + \dim \text{Tr}^{-1}(0) .$$

Thus the dimension of $\text{Tr}^{-1}(0)$ is just one less than $\dim u(n)$ i.e. $n^2 - 1$. But $su(n) = \text{Tr}^{-1}(0)$ is just the tangent space of $SU(n)$, since

$$\text{Tr}(C) = 0 \Leftrightarrow 1 = e^{\text{Tr}(C)} = \det e^C$$

$$\Leftrightarrow e^C \in SU(n) .$$

D. <u>Lie algebras</u>

It is easy to see that $so(n)$, $su(n)$ and $sp(n)$ are not closed under matrix multiplication. For example, if

$$\alpha = \begin{pmatrix} 0 & x \\ -x & 0 \end{pmatrix} \quad \text{then} \quad \alpha^2 = \begin{pmatrix} -x^2 & 0 \\ 0 & -x^2 \end{pmatrix}$$

which is not skew-symmetric.

<u>Proposition</u> 10: <u>For</u> $k \in \{\mathbb{R}, \mathbb{C}, \mathbb{H}\}$ <u>and</u> $A, B \in M_n(k)$ <u>we define</u>

$$[A, B] = AB - BA .$$

Then so(n) , su(n) and sp(n) are closed under [,] .

Proof: We need to show that

$$(AB - BA) + {}^t\overline{(AB - BA)} = 0 .$$

The left-hand side is

$$AB-BA+{}^t\overline{B}\,{}^t\overline{A}-{}^t\overline{A}\,{}^t\overline{B} = AB+({}^t\overline{A}\overline{B}-{}^t\overline{A}\overline{B})-BA+(-{}^t\overline{B}\overline{A}+{}^t\overline{B}\overline{A})+{}^t\overline{B}\,{}^t\overline{A}-{}^t\overline{A}\,{}^t\overline{B}$$

$$= A(B+{}^t\overline{B})-(A+{}^t\overline{A})\,{}^t\overline{B}-B(A+{}^t\overline{A})+(B+{}^t\overline{B})\,{}^t\overline{A}$$

$$= 0 .$$

Thus so(n) , su(n) and sp(n) become algebras (over R) with this
bracket multiplication. This product has some obvious properties.

    (i) $[A,B] = [B,A]$

   (ii) $[A,B+C] = [A,B] + [A,C]$

        $[A+B,C] = [A,C] + [B,C]$

  (iii) For $r \in R$, $r[A,B] = [rA,B] = [A,rB]$.

Finally, [,] has one nonobvious property.

   (iv) $[A,[B,C]] + [B,[A,C]] + [C,[A,B]] = 0$.

Property (iv) is called the Jacobi identity and its proof is a routine
verification.

Definition: A real vector space with a product satisfying
(i)...(iv) is called a Lie algebra. (One could clearly consider complex Lie algebras, but we will have no occasion to do so.)

Let us consider low dimensional Lie algebras. For dim 1 the
vector space is just R and if x,y $\in$ R we have

$$[x,y] = x[1,y] = xy[1,1] = 0 \quad (\text{by (i)}) \ .$$

So we have the trivial product (which obviously satisfies (i)...(iv)).

Consider R^2 with basis e_1, e_2

We must have

$$[e_1, e_1] = 0 \ , \ [e_2, e_2] = 0 \quad \text{and} \quad [e_1, e_2] = -[e_2, e_1] \ .$$

Let $[e_1, e_2] = ae_1 + be_2$. Then, for example,

$$[e_1, [e_1, e_2]] = [e_1, (ae_1 + be_2)] = a[e_1, e_1) + b(e_1, e_2]$$

$$= b(ae_1 + be_2) \ .$$

By the Jacobi identity

$$[e_1, [e_1, e_2]] + [e_1, [e_2, e_1]] + [e_2, [e_1, e_1]] = 0 \ ,$$

so

$$b(ae_1 + be_2) + [e_1, (-ae_1 - be_2)] = 0$$

which is true with no conditions on a, b . If we take $a = 0 = b$ we get the trivial Lie algebra. For any other choice we get a non-trivial Lie algebra. In the exercises one shows that these nontrivial 2-dimensional Lie algebras are all "essentially the same."

We will not try to find out all nontrivial 3-dimensional Lie algebras, but will simply look at two which arise quite naturally.

$$so(3) = \left\{ \begin{pmatrix} 0 & a & b \\ -a & 0 & c \\ -b & -c & 0 \end{pmatrix} \middle| a, b, c \in R \right\}$$

clearly has dimension three. Also using basis

and defining

$$[i,j] = k$$
$$[j,k] = i$$
$$[k,i] = j$$

gives a 3-dimensional Lie algebra.

E. Exercises

1. Let A be a 3×3 skew-symmetric matrix. Show that A^2 is symmetric, but show by example that A^3 could be neither symmetric nor skew-symmetric.

2. Let $B \in \mathfrak{o}(3) - SO(3)$. Show that the series for $\log B$ does not converge.

3. Prove the Jacobi identity for $[A,B] = AB - BA$.

4. Prove that any two nontrivial Lie algebras on R^2 are isomorphic as Lie algebras.

5. Show that the two 3-dimensional Lie algebras defined above (§D) are isomorphic.

Chapter 5
SO(3) and Sp(1)

A. The homomorphism $\rho : S^3 \to SO(3)$

We have seen that $Sp(1)$, which is all quaternions of unit length, is just the unit 3-sphere in R^4 (= $\mathbb{H}$). Also we have seen that $\dim SO(3) = \frac{3 \cdot 2}{2} = 3$. So dimension won't distinguish S^3 from $SO(3)$, and, for all we know now, they might be isomorphic. In this section we define and study an "almost isomorphism" between them.

Proposition 1: If $q \in S^3$, then the "left translation"

$$L_q : \mathbb{H} \to \mathbb{H}$$

given by $L_q(q') = qq'$ is an orthogonal map of R^4 to R^4 .

Proof: As vector spaces over R , $\mathbb{H}$ and R^4 are the same. So L_q is surely a linear map of R^4 , for if $a,b \in R$ and $\alpha, \beta \in \mathbb{H}$ we have

$$L_q(a\alpha + b\beta) = q(a\alpha + b\beta) = aq\alpha + bq\beta = aL_q(\alpha) + bL_q(\beta) .$$

To see that L_q is orthogonal, it suffices to show that L_q preserves the perpendicularity (using $\langle \, , \, \rangle$ for R^4) of the four unit vectors $1,i,j,k$. For example, let $q = a + ib + jc + kd$ and

calculate $\langle L_q(i), L_q(j) \rangle$ (using $\langle \ , \ \rangle$ for R^4) . We get
$ad + bc - bc - da = 0$. For $\langle L_q(1), L_q(i) \rangle$ we get

$$\langle a + ib + jc + kd, \ ai - b - kc + jd \rangle = -ab + ab + dc - dc = 0 \ .$$

The computations for other pairs of basis vectors are similar.

Definition of ρ: For $q \in S^3$ and $\alpha \in \mathbb{H}$ we define

$$\rho(q)(\alpha) \quad q\alpha\bar{q} \ .$$

That is, we do a left translation by q and a right translation by
$\bar{q}$. By Proposition 1 this is an orthogonal map of R^4 to R^4 ; i.e.,
$\rho(q) \in \mathbb{O}(4)$.

Since real quaternions commute with all other quaternions, if
x is a real quaternion

$$\rho(q)x = qx\bar{q} = xq\bar{q} = x \ .$$

Note also that $\rho(\bar{q})$ is the inverse of $\rho(q)$ in the group $\mathbb{O}(4)$
since $\rho(q)\rho(q^{-1})(\alpha) = q(\bar{q}\alpha q)\bar{q} = \alpha$ and similarly for $\rho(\bar{q})\rho(q)$.
Together these two observations imply that $\rho(q)$ maps the 3-space
spanned by i,j,k to itself (Exercise #3). Thus $\rho(q)$ can be considered as an element of $\mathbb{O}(3)$ (Exercise #4).

Fact (to be proved after Chapter VI): $\rho(q)$ is in $S\mathbb{O}(3)$.

Proposition 2: $\rho : S^3 \to S\mathbb{O}(3)$ is a surjective homomorphism and

$$\text{Ker}(\rho) = \{1, -1\} \subset S^3 \ .$$

Proof: If $q_1, q_2 \in S^3$ and $\alpha \in \text{Span}\langle i, j, k \rangle$, then

$$\rho(q_1 q_2)(\alpha) = q_1 q_2 \alpha \overline{q_1 q_2} = q_1(q_2 \alpha \overline{q_2})\bar{q} = \rho(q_1)\rho(q_2)(\alpha) \ .$$

Thus ρ is a homomorphism.

Clearly $\rho(1)$ and $\rho(-1)$ are the identity in $SO(3)$ so that 1 and -1 are in Ker ρ . Conversely, suppose $\rho(q)$ is the identity with $q = a + ib + jc + kd$. Then $\rho(q)(i) = i$ gives $(a + ib + jc + kd)(i)(a - ib - jc - kd) = i$. And from this we get $a^2 + b^2 - c^2 - d^2 = 1$. But $a^2 + b^2 + c^2 + d^2 = 1$ and we conclude that $c = 0 = d$. From $\rho(q)j = j$ we get $b = 0$. Then $a^2 = 1$ so $a \in \{1,-1\}$.

Finally we need to show that ρ is surjective. This will be quite easy once we know some topology (Chapter VI)--otherwise it is an almost hopelessly complicated computation. Here we will just show that we can find a $q \in S^3$ such that $\rho(q)$ is the element of $SO(3)$ which leaves k fixed, sends i to j and sends j to $-i$.

Let $q = a + ib + jc + kd$. We want
$(a + ib + jc + kd)(k)(a - ib - jc - kd) = k$, or
$(a + ib + jc + kd)(ka - jb + ic + d) = k$, so

$$ad - bc + bc - ad = 0 \quad \text{(automatically)} \ ,$$

$$ac + bd + ac + bd = 0 \quad \text{or} \quad 2(ac + bd) = 0 \ ,$$

$$-ab + cd + dc - ab \quad \text{or} \quad 2(cd - ab) = 0 \ .$$

$$a^2 + d^2 - b^2 - c^2 = 1 \ .$$

Now

$$a^2 + d^2 + b^2 + c^2 = 1$$

so

$$2(b^2 + c^2) = 0 \quad \text{or} \quad b = 0 = c.$$

So the only condition on q such that $\rho(q)k = k$ is that $q = a + dk$ (with $a^2 + d^2 = 1$).

Next we want $\rho(q)i = j$.

$$(a + kd)(i)(a - kd) = j$$

$$(a + kd)(ia + jd) = j$$

$$a^2 - d^2 = 0 , \quad a = \pm d$$

$$ad + ad = 1 , \quad \text{if} \quad a = d , \quad 2a^2 = 1 ,$$

and we can't have $a = -d$. Finally we insist that $\rho(q)j = -i$. So

$$(a + ka)(j)(a - ka) = -i$$

$$(a + ka)(ja - ia) = -1$$

$$-2a^2 = -1$$

$$a^2 - a^2 = 0.$$

So $q = \dfrac{1}{\sqrt{2}} + k\dfrac{1}{\sqrt{2}}$ or $q = -\dfrac{1}{\sqrt{2}} - k\dfrac{1}{\sqrt{2}}$. Both will give the desired element of $SO(3)$. (This should be enough to convice us that we should not try the general proof of surjectivity at this stage.)

Note that this does not prove that S^3 and $SO(3)$ are not isomorphic. ρ is not an isomorphism, but one might exist. In the next section we give a fairly easy proof that $S^3 \neq SO(3)$.

B. Centers

In Exercise #4 of Chapter I the center C of a group G is defined as

$$C = \{x \in G \mid xy = yx \text{ for all } y \in G\} ,$$

and was shown to be an abelian and normal subgroup of G . We leave
it as an exercise here to show that any isomorphisms of groups induces
an isomorphism of their centers. We will show that $S^3 \not\cong SO(3)$ by
showing that their centers are not isomorphic.

Proposition 3: The center of $S^3 = Sp(1)$ is $\{1,-1\}$, whereas
the center of SO(3) is $\{I\}$.

Proof: Since real quaternions commute with all quaternions, it
is clear that $\{1,-1\} \subset$ Center S^3 . Conversely, suppose
$q = a + ib + jc + kd \in S^3$ is in the center. Then $qi = iq$ gives

$$ai - b - ck + dj = ai - b + ck - dj$$

so that $c = 0 = d$. Then $qj = jq$ gives

$$(a + ib)j = j(a + ib)$$

and this implies $b = 0$. So $q = a$ and $a^2 = 1$. Thus Center S^3
$= \{1,-1\}$.

Suppose $A \in SO(3)$ is in the center. Since A commutes with
all elements of SO(3) it surely commutes with all elements of

$$T = \begin{pmatrix} \cos\theta & \sin\theta & 0 \\ -\sin\theta & \cos\theta & 0 \\ 0 & 0 & 1 \end{pmatrix}$$

since $T \subset SO(3)$. Consider the standard basis $e_1 = (1,0,0)$, $e_2 = (0,1,0)$, $e_3 = (0,0,1)$ for R^3 .

Claim: A leaves e_3 fixed (or sends it to $-e_3$) . Choose $B \in T$ which sends e_1 to e_2 and e_2 to $-e_1$ and (automatically) leaves e_3 fixed. Then set $Ae_3 = ae_1 + be_2 + ce_3$. Then $BAe_3 = ae_2 - be_1 + ce_3$, whereas $ABe_3 = Ae_3$; this implies $a = 0 = b$ and since A preserves length, we must have $c = 1$, or $c = -1$.

Thus A induces an orthogonal map of the $e_1 e_2$ plane. Actually, it is a rotation because:

Sublemma: Any element of $\mathcal{O}(2)$ which commutes with all rotations, is itself a rotation.

Let $\phi : R^2 \to R^2$ denote such an element of $\mathcal{O}(2)$. For any rotation

$$t = \begin{pmatrix} \cos\theta & \sin\theta \\ -\sin\theta & \cos\theta \end{pmatrix}$$

we must have $\phi t = t\phi$. Let $\phi = \begin{pmatrix} \alpha & \beta \\ \gamma & \delta \end{pmatrix}$. We get

$$\alpha\cos\theta - \beta\cos\theta = \alpha\cos\theta + \gamma\sin\theta$$

$$\alpha\sin\theta + \beta\cos\theta = \beta\cos\theta + \delta\sin\theta ,$$

holding for all θ . So $\gamma = -\beta$ and $\alpha = \delta$. Thus

$$\phi = \begin{pmatrix} \alpha & \beta \\ -\beta & \alpha \end{pmatrix} \quad \text{and} \quad \det\phi = \alpha^2 + \beta^2 .$$

66

Since this cannot equal -1 (and must be in {1,-1}),   this proves the sublemma.

This also proves that c = 1 (not -1) (since A ∈ SO(3)) and we conclude that

$$A \in T .$$

We can now finish the proof.

$$A = \begin{pmatrix} \cos\theta & \sin\theta & 0 \\ -\sin\theta & \cos\theta & 0 \\ 0 & 0 & 1 \end{pmatrix}$$

and we let

$$R = \begin{pmatrix} 0 & 0 & 1 \\ 0 & 1 & 0 \\ -1 & 0 & 0 \end{pmatrix} \in SO(3) .$$

Since A must commute with R we get

$$AR = \begin{pmatrix} 0 & \sin\theta & \cos\theta \\ 0 & \cos\theta & -\sin\theta \\ -1 & 0 & 0 \end{pmatrix} = \begin{pmatrix} 0 & 0 & 1 \\ -\sin\theta & \cos\theta & 0 \\ -\cos\theta & \sin\theta & 0 \end{pmatrix} = RA .$$

Thus we must have $\cos\theta = 1$ and $\sin\theta = 0$. Thus A = I and Proposition 3 is proved.

We will calculate the centers of all of the groups SO(n) , U(n) , SU(n) , Sp(n) in a later chapter, after we know about maximal tori. We conclude this chapter with a bit more abstract theory which we will need later.

C. Quotient groups

If H is a subgroup of G we define an equivalence relation $\sim$ on G by

$$x \sim y \quad \text{if} \quad xy^{-1} \in H .$$

This relation is <u>reflexive</u>, $x \sim x$, since $xx^{-1} = e \in H$. It is <u>symmetric</u>, $x \sim y \Rightarrow y \sim x$, since $xy^{-1} \in H \Rightarrow (xy^{-1})^{-1} = yx^{-1} \in H$. It is <u>transitive</u> , $x \sim y$ and $y \sim z \Rightarrow x \sim z$, since $xy^{-1} \in H$ and $yz^{-1} \in H$ imply that $(xy^{-1})(yz^{-1}) = xz^{-1} \in H$. Thus $\sim$ divides G into equivalence classes.

Let $C(x)$ denote the class containing x . Then

$$C(x) = Hx = \{hx \mid h \in H\} .$$

Also $Hx = Hy \Leftrightarrow xy^{-1} \in H \Leftrightarrow y \in C(x) \Leftrightarrow x \in C(y)$. These equivalence classes are called <u>right cosets</u> of H .

Example: Let $G = S^3$ ($= Sp(1)$) and $H = \{1,-1\}$. Then $Hq = \{q,-q\} = H(-q)$ so each equivalence class contains exactly two points of S^3 .

Example: In $G = U(3)$ let

$$H = \{\lambda I \mid \lambda \text{ a complex number of unit length}\} .$$

Then H is a circle subgroup of G and the right cosets are circles. Thus $U(3)$ can be divided into disjoint circles filling up $U(3)$.

Similarly, let $G = S^3 = Sp(1)$ and let H be the circle $\{a + ib \mid a^2 + b^2 = 1\}$. Thus S^3 can be divided up into circles.

One defines left cosets in a similar manner

$$xH = \{xh \mid h \in H\} \ .$$

Recall (Exercise #3, Chapter I) that a subgroup H is <u>normal</u> if $xHx^{-1} = H$ for all x in G .

Observation: A subgroup H of G is normal (in G) ⇔ xH = Hx for every x ∈ G .

Let G/H denote the set whose elements are the rightcosets of H in G .

<u>Proposition</u> 4: <u>If</u> H <u>is a normal</u> <u>subgroup</u> <u>of</u> G , <u>then</u> <u>the</u> <u>operation</u> <u>on</u> G/H <u>defined</u> <u>by</u>

$$(Hx)(Hy) = H(xy)$$

<u>makes</u> G/H <u>into</u> <u>a</u> <u>group.</u>

Proof: We need H normal to show the operation on G/H is well defined. Suppose Hx = Hz and Hy = Hw . We must show that Hxy = Hzw . Well, $xy(zw)^{-1} = xyw^{-1}z^{-1}$ and $yw^{-1} = h_1 \in H$. Also, $z^{-1} = x^{-1}h_2$. So

$$xy(zw)^{-1} = xh_1x^{-1}h_2$$

and, since H is normal, $xh_1x^{-1} = h_3 \in H$ so that

$$xy(zw)^{-1} = h_3h_2 \in H \ ,$$

and we have proved that the operation is well defined.

The rest is easy. $H = He \in G/H$ is the identity and Hx^{-1} is the inverse of Hx . (Associativity is inherited from G - (Hx)(HyHz) = (HxHy)Hz since x(yz) = (xy)z) .

Example: $G = Sp(1)$ and $H = \{1,-1\}$. H is the center of G and thus is a normal subgroup. Thus G/H is a group. We know it is $SO(3)$.

There is a natural map $\eta : G \to G/H$ given by $\eta(x) = Hx$. In the exercises it is shown that η is a surjective homomorphism with kernel H.

Let G be a group and $x,y \in G$. Then the element

$$xyx^{-1}y^{-1}$$

is called the commutator of x and y (because $(xyx^{-1}y^{-1})(yx) = xy$). Now the product of two commutators is not necessarily a commutator, but we set $[G,G] = \{$all finite products of commutators$\}$.

Proposition 5: $[G,G]$ is a normal subgroup of G and $\dfrac{G}{[G,G]}$ is an abelian group.

Proof: Closure and identity are clear and

$$(xyx^{-1}y^{-1})(yxy^{-1}x^{-1}) = e \ ,$$

showing $[G,G]$ is a subgroup. Let $z \in G$ and $xyx^{-1}y^{-1} \in [G,G]$. Then

$$z(xyx^{-1}y^{-1})z^{-1} = zxy(z^{-1}(xy)^{-1}(xy)z)x^{-1}((yz)^{-1}(yz))y^{-1}z^{-1}$$

$$= \{z(xy)z^{-1}(xy)^{-1}\}\{x(yz)x^{-1}(yz)^{-1}\}\{yzy^{-1}z^{-1}\} \in [G,G] \ .$$

This easily extends to products of commutators, so that $[G,G]$ is a normal subgroup.

Finally, $[G,G]x[G,G]y = [G,G]xy = [G,G]yx = [G,G]y[G,G]x$ since

$$xy(yx)^{-1} = xyx^{-1}y^{-1} \in [G,G] .$$

q.e.d.

In most instances we will encounter, if G is a matrix group and C is its center, then G/C will have trivial center. But this need not always be the case.

Proposition 6: For $x \in G$ define

$$\psi(x) : G \to G$$

by $\psi(x)(y) = xyx^{-1}y^{-1}$. Then G/C has nontrivial center $\Leftrightarrow \exists x \in G-C$ such that

$$\psi(x)(G) \subset C .$$

Proof: $\Leftarrow$

$x \notin C \Rightarrow Cx \neq C$ so Cx is not the identity in G/C . But for any $y \in G$ we have $xyx^{-1}y^{-1} \in C$ so that $CxCy = Cxy = Cyx = CyCx$ and $Cx \in$ center G/C .

$\Rightarrow$

Conversely, $Cx \neq C$ with Cx in the center implies

$CxCy = Cxy = Cyx = CyCx$ so that $xyx^{-1}y^{-1} \in C$ for all $y \in G$.

Once we have done a little topology (Chapter VI) we easily have:

Corollary: If G is connected and C is discrete (in particular if C is finite), then G/C has no center.

D. Exercises

1. Let G be a group and $x \in G$. Show that left translation $L_x : G \to G$ by x ($L_x(g) = xg$) is a one-to-one map of G onto G. Let R_x be right translation so that

$$R_{x^{-1}} \circ L_x(g) = xgx^{-1} .$$

Show that $R_{x^{-1}} \circ L_x$ is an isomorphism of G onto G.

2. Do one more step in the proof of Proposition 1 by showing $\langle L_q(i), L_q(k) \rangle = 0$.

3,4. These are listed by number in the text.

5. Show that $\rho(i)$, $\rho(j)$, $\rho(k)$ are all in $SO(3)$.

6. Show that the set T defined in the proof of Proposition 3 is an abelian subgroup of $SO(3)$.

7. Let $\phi : G \to K$ be a surjective homomorphism of groups and $H = \text{Ker } \phi$. Then we have

$$G \overset{\phi}{\to} K$$
$$\downarrow \eta \qquad .$$
$$G/H$$

Show that $\phi \circ \eta^{-1}$ is well defined and gives an isomorphism of G/H onto K.

8. Show that (see Exercise #6) the abelian subgroup T of $SO(3)$ is not a normal subgroup.

9. Show that the subgroup $H = \{a + ib \mid a^2 + b^2 = 1\}$ of $Sp(1)$

is not a normal subgroup.

10. Show an isomorphism of groups induces an isomorphism of their centers.

Chapter 6
Topology

A. Introduction

Our matrix groups are all subsets of euclidean spaces, because
they are all subsets of

$$M_n(\mathbb{R}) = \mathbb{R}^{n^2} \quad \text{or} \quad M_n(\mathbb{C}) = \mathbb{R}^{2n^2} \quad \text{or} \quad M_n(\mathbb{H}) = \mathbb{R}^{4n^2} \, .$$

There are certain topological properties, notably connectedness and
compactness, which some of our groups have and others do not. These
properties are preserved by continuous maps and so are surely invariants under isomorphisms of groups. So a connected matrix group could
not be isomorphic with a nonconnected matrix group, and a similar
statement holds for compactness. We will define these properties and
decide which of our groups have them. This will be done in sections
B and C.

In section D we define and discuss the notion of a countable basis
for open sets, a concept we will need in our study of maximal tori in
matrix groups. Finally, in section E we define manifold and show that
all of our matrix groups are manifolds. Then we prove a theorem
about manifolds which gives an easy proof that the homomorphism
$\rho : Sp(1) \to SO(3)$ (defined in Chapter V) is surjective.

B. Continuity of functions, open sets, closed sets

Definition: A metric d on a set S is a way of assigning to each $x, y \in S$ a real number $d(x,y)$ (the distance from x to y) in such a way that:

(i) $d(x,y) \geq 0$ and $d(x,y) = 0 \Leftrightarrow x = y$,

(ii) $d(x,y) = d(y,x)$,

(iii) $d(x,y) + d(y,z) \geq d(x,z)$.

Condition (iii) is called the triangle inequality.

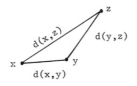

We will define such a metric d on R^n and then for any $S \subset R^n$ d will also clearly be a metric on S . Recall that for $x, y \in R^n$,

$$x = (x_1, \ldots, x_n) \qquad y = (y_1, \ldots, y_n)$$

we defined an inner product

$$\langle x, y \rangle = x_1 y_1 + \ldots + x_n y_n .$$

Set $d(x,y) = \sqrt{\langle x-y, x-y \rangle}$. (Thus we define $d(x,y)$ to be the length of the vector x - y .)

Proposition 1: This is a metric on R^n .

Proof: Properties (i) and (ii) follow from

$$\langle x, x \rangle \geq 0 \quad \text{and} \quad \langle x, x \rangle = 0 \Leftrightarrow x = 0$$

and symmetry of the inner product. To prove the triangle inequality

we will prove the corresponding property of $\langle\ ,\ \rangle$ called the

Schwarz inequality.

For any $x,y \in \mathbb{R}^n$ and $t \in \mathbb{R}$ we have

$$\langle x+ty, x+ty \rangle \geq 0 \ .$$

Using the bilinearity and symmetry of $\langle\ ,\ \rangle$ this gives

$$\langle x,x \rangle + 2\langle x,y \rangle t + \langle y,y \rangle t^2 \geq 0 \ .$$

This quadratic polynomial in t with real coefficients is always

≥ 0 and thus it cannot have two distinct real roots. (A quadratic

polynomial can have only one minimum.) Thus the discriminant cannot

be positive; i.e.,

$$(2\langle x,y \rangle)^2 - 4\langle y,y \rangle \langle x,x \rangle \leq 0 \ .$$

So

(*) $$\langle x,y \rangle^2 \leq \langle x,x \rangle \langle y,y \rangle \ .$$

The inequality (*) is the Schwarz inequality.

We apply (*) to the vectors x - y and y - z to get

(**) $$\langle x-y, y-z \rangle \leq \sqrt{\langle x-y, x-y \rangle}\ \sqrt{\langle y-z, y-z \rangle} \ .$$

If we square both sides of (iii), write it in terms of $\langle\ ,\ \rangle$ and use

basic properties of $\langle\ ,\ \rangle$, we see that (iii) is equivalent to (**).

We use this metric d on $\mathbb{R}^n$ to define open balls. Let $x \in \mathbb{R}^n$

and $r > 0$ be a real number. Set

$$B(x,r) = \{y \in \mathbb{R}^n \mid d(x,y) < r\}$$

and call this the open ball with center x and radius r . Open balls

in euclidean spaces allow us to give a fairly direct generalization of

the notion of continuity of a function on R to functions defined on

spaces of dimension greater than one.

Let A be a subset of R^n and

$$f : A \to R^m$$

be a function defined on A and taking values in some euclidean space
R^m .

Definition: To say f <u>is</u> <u>continuous</u> <u>at a</u> <u>point</u> a ∈ A means:

Given any open ball B(f(a),ε) in R^m , there exists an open

ball B(a,δ) in R^n such that any point x ∈ A ∩ B(a,δ) satisfies

$$f(x) \in B(f(a),\varepsilon) .$$

Another way of saying this is: Given ε > 0   there exists δ > 0

such that if x ∈ A satisfies d(a,x) < δ , then f(x) satisfies

d(f(a),f(x)) < ε .

Both ways are just precise ways of saying that f is continuous

if it sends "nearby points" of A to "nearby points" in R^m .

It is important to notice that the continuity of f depends on

the domain A of definition. For example define

$$f : R \to R$$

by

$$f(x) = \begin{cases} 0 & \text{if } x < 0 \\ 1 & \text{if } x \ge 0 \end{cases} .$$

Then f is not continuous at 0 . But suppose we take A ⊂ R to be all $x \geq 0$ and restrict f to A

$$f : A \to R .$$

Then this restricted f is continuous at 0 .

A function $f : A \to R^m$ $(A \subset R^n)$ is said to be <u>continuous</u> if it is continuous at each $a \in A$.

<u>Example</u>: If $A \subset R^n$ is a finite set then any $f : A \to R^m$ is continuous.

<u>Proof</u>: For any $a \in A$ let $b_1, \ldots, b_k$ be all of the other points of A . Let

$$\delta_i = d(a, b_i) , \quad i = i, \ldots, k$$

and let δ be the smallest of these. Then for any $\varepsilon > 0$ any element of A in $B(a, \delta)$ goes into $B(f(a), \varepsilon)$ (because a is the only such element of A) .

<u>Proposition</u> 2: <u>If</u> $A \subset R^n$ <u>and</u>

$$A \overset{f}{\to} R^m , \ f(A) \overset{g}{\to} R^p$$

<u>are continuous, then</u> $g \circ f$ <u>is continuous.</u>

<u>Proof</u>: Let $a \in A$ and $\varepsilon > 0$. Since g is continuous, there exists $\eta > 0$ such that every element of f(A) in $B(f(a), \eta)$ is sent by g into $B(g(f(a)), \varepsilon)$. Since f is continuous there exists $\delta > 0$ such that every element of A in $B(a, \delta)$ is sent by f into $B(f(a), \eta)$ and is then sent by g into $B(g(f(a)), \varepsilon)$.

q.e.d.

Some exercises on continuity are given at the end of this chapter.

Definition: A set $U \subset R^n$ is an open set if each $x \in U$ lies in some $B(x,r) \subset U$ (where r will depend on x).

Clearly R^n is an open set. It is not quite so clear that the empty set ϕ is open; but since there is no $x \in \phi$, there is no requirement that some $B(x,r)$ be contained in ϕ.

Example: Any open ball $B(y,s)$ is an open set. Let $x \in B(y,s)$ (i.e., $d(x,y) < s$). We must find $r > 0$ such that $B(x,r) \subset B(y,s)$. Well, let $r = s - d(x,y)$. If $z \in B(x,r)$ then $d(z,x) < s - d(x,y)$

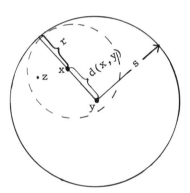

and thus

$$d(z,x) + d(z,y) < s .$$

Using the triangle inequality we have

$$d(z,y) \leq d(z,x) + d(x,y) < s$$

showing $z \in B(y,s)$.

Example: $(0,1) = \{x \in R \mid 0 < x < 1\}$ is an open set in R (because it is the open ball $B(\tfrac{1}{2},\tfrac{1}{2})$). But

$(0,1] = \{x \in R \mid 0 < x \le 1\}$ is not open in R because $1 \in (0,1]$ but no $B(1,r)$ lies in $(0,1]$ since every such ball contains numbers greater than 1 .

Definition: A subset $C \subset R^n$ is defined to be <u>closed</u> if its complement $R^n - C$ is open.

$(0,1] \subset R$ is neither open nor closed. We have seen it is not open. Let $T = R - (0,1]$. Then $0 \in T$ but no $B(0,r)$ can lie in T since each will contain points of $(0,1]$. Thus T is not open so that $(0,1]$ is not closed.

$[0,1] = \{x \in R \mid 0 \le x \le 1\}$ is a closed set.

Example: Let $K \subset R^n$ be a finite set. Then K is closed.

For if $x \in R^n - K$ there will be a minimum distance δ from x to points of K . Then

$$B(x,\delta) \subset R^n - K \, ,$$

proving K is closed.

C. <u>Connected</u> <u>sets</u>, <u>compact</u> <u>sets</u>

Definition: A set D in R^n is <u>connected</u> if: Given $x,y \in D$ there exists a continuous function

$$\gamma : [0,1] \to D$$

(i.e., $\gamma : [0,1] \to R^n$ with $\gamma([0,1]) \subset D$)

with $\gamma(0) = x$ and $\gamma(1) = y$.

Such a function may be called a path from x to y in D .

Examples: R^n is connected because

$$\gamma(t) = (x + t(y-x))$$

is a path in R^n from x to y .

  $D = \{x \in R \mid x \neq 0\}$ is not connected. For example, no path
from -1 to 1 in R can lie in D .

  $D = \{(x_1, x_2) \in R^2 \mid (x_1, x_2) \neq (0,0)\}$ is connected.

Important example: $\Theta(n) \subset R^{n^2}$ is not connected. The matrices

$$A = \begin{pmatrix} -1 & & & \bigcirc \\ & 1 & & \\ & & \ddots & \\ \bigcirc & & & 1 \end{pmatrix} \quad \text{and} \quad I = \begin{pmatrix} 1 & & & \bigcirc \\ & 1 & & \\ & & \ddots & \\ \bigcirc & & & 1 \end{pmatrix}$$

are in $\Theta(n)$. If $\gamma : [0,1]$
$\rightarrow \Theta(n)$ is a path from A to I , then the composite function

$$[0,1] \xrightarrow{\gamma} \Theta(n) \xrightarrow{\det} R$$

would be continuous (Proposition 2) . But it would be a path in
$\{-1,1\} \subset R$ from -1 to 1 , contradicting the existence of γ .

  Recall that $so(n) \subset M_n(R)$ consists of skew-symmetric matrices
and that if $A \in so(n)$, then $\exp A \subset \Theta(n)$.

Proposition 3: exp maps so(n) into SO(n) .

Proof: For $B \in so(n)$ the path

$$\gamma(t) = e^{tB}$$

is a path from $e^0 = I$ to e^B . As seen above, this implies that
$\det e^B = +1$ so $e^B \in SO(n)$.

Proposition 4: Let $D \subset R^n$ be connected and

$$f : D \to R^m$$

be continuous, then $f(D)$ is connected.

Proof: Given $a, b \in f(D)$, choose $x, y \in D$ such that $f(x) = a$ and $f(y) = b$. Choose a path γ from x to y in D . Then $f \circ \gamma$ is a path from a to b in $f(D)$.

Definition: A subset W of R^n is bounded if W lies in some open ball. This is clearly equivalent to: W lies in some $B(0, r)$.

Now boundedness, unlike connectedness, is not preserved by continuous functions. For example, if $W = (0, 1) \subset R$ and $f : W \to R$ is defined by $f(x) = \frac{1}{x}$, then W is bounded but $f(W)$ is not bounded.

Neither is the property of being closed preserved by continuous functions. For example, R is closed in R and $f : R \to R$ defined by $f(x) = e^x$ is continuous, but $f(R) = \{y \in R \mid y > 0\}$ is not closed.

However, when we put closed and bounded together, they are then both preserved.

Definition: $C \subset R^n$ is compact if it is closed and bounded.

Proposition 5: If $C \subset R^n$ is compact and

$$f : C \to R^m$$

is continuous, then $f(C)$ is compact.

The proof is relegated to an appendix.

D.   Subspace topology, countable bases

Sometimes we will have a subset W of R^n and will want to
know which subsets of W we should call open sets in W .

Definition: If $U \subset W \subset R^n$ we say U is an open set in W
if there exists an open set V in R^n such that

$$U = V \cap W .$$

For example, if $W = [0,1] \subset R$, then
$U = (\frac{1}{2},1] = \{x \in R \mid \frac{1}{2} < x \leq 1\}$ is an open set in W , but not an
open set in R . $U' = [\frac{1}{2},1]$ is not open in W .

Note that if W is an open set in R^n, then $U \subset W$ is open in
W if and only if it is open in R^n .

For $W \subset R^n$ the collection of all open sets of W is the
subspace topology of W .

Recall that $V \subset R^n$ is defined to be open if any $x \in V$ has
some $B(x,r) \subset V$. This is equivalent to saying that $V \subset R^n$ is open
if it is either the empty set or is a union of open balls. (Exercise.
Of course, not every open set is an open ball, but open balls suffice
to give all open sets by taking unions (the "empty" union being
allowed).

Definition: A collection $\upsilon = \{V_\alpha\}$ of open sets in R^n is a
basis for open sets if every open set in R^n is a union of some of
the V_α's .

Examples: The set of all open squares in R^2 is a basis for the open sets in R^2.

The set of all open intervals $(a,b) \subset R$ with a and b rational is a basis for the open sets in R.

The set of all open balls in R^n is, of course, a basis for open sets. But so is

$$\{B(x,r) \mid x = (x_1,\ldots,x_n) \text{ with each } x_i \text{ rational, and } r \text{ is rational}\}$$

(See Proposition 7.)

For a subset W of R^n we know which are the open sets in W and we can give the same definition as above for the notion of a basis for the open sets in W. Indeed, it is clear that if $\upsilon = \{V_\alpha\}$ is a basis for the open sets in R^n, then $\{V_\alpha \cap W\}$ is a basis for the open sets of W.

We want to get bases for open sets which are "minimial" in the sense that they have no more sets than needed to do the job. The notion that comes up is countability.

Definition: A set S is countable if its elements can all be arranged in a finite or infinite sequence $s_1, s_2, s_3, \ldots$; that is, every element of S will be somewhere in the sequence.

Examples: The set $\mathbb{Q}$ of all positive rational numbers is countable; for example

$$1,\ 2,\ \frac{1}{2},\ \frac{3}{2},\ 3,\ \frac{1}{3},\ \frac{2}{3},\ \frac{4}{3},\ \frac{5}{3},\ \frac{7}{3},\ \frac{8}{3},\ 4,\ \frac{1}{4},\ \ldots$$

is a sequence containing all positive rationals. Similarly,

1, -1, 2, -2, $\frac{1}{2}$, $-\frac{1}{2}$, $\frac{3}{2}$, $-\frac{3}{2}$, 3, -3,... contains all of $\mathbb{Q}$.

The set $I = [0,1] = \{x \in \mathbb{R} \mid 0 \le x \le 1\}$ is not countable. We prove this contrapositively. Suppose

$$r_1, r_2, r_3, \ldots$$

is a list of all elements of I . It suffices to give an element of I which cannot be in the list. Express the r_i's as decimals

$$r_i = .x_{i_1} x_{i_2} x_{i_3} \ldots \quad .$$

Let $r = .y_1 y_2 y_3 \ldots$ where $y_j = 5$ if $x_{jj} \ne 5$ and $y_j = 1$ if $x_{jj} = 5$. Then

$$r \ne r_1 \quad \text{because} \quad y_1 \ne x_{11}$$

$$r \ne r_2 \quad \text{because} \quad y_2 \ne x_{22} , \quad \text{etc.}$$

But $r \in I$.

Proposition 6: If A and B are countable sets, then so is their cartesian product $A \times B$.

Proof: Let $A = \{a_1, a_2, a_3, \ldots\}$ and $B = \{b_1 b_2, b_3, \ldots\}$. Then we can write

$$A \times B = \{(a_1, b_1), (a_1, b_2), (a_2, b_1), (a_3, b_1), (a_2, b_2), \ldots \}$$

because following the path shown below will include all of $A \times B$.

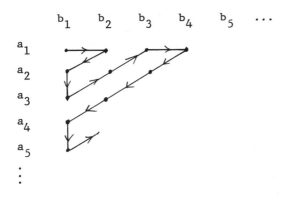

Proposition 7: R^n (and hence any $W \subset R^n$) has a countable basis for its open sets.

Proof: The set

$$C = \{B(x,r) \mid x = (x_1, \ldots, x_n) \text{ each } x_i \in \mathbb{Q} \text{ and } r \in \mathbb{Q}\}$$

can be put in 1-1 correspondence with (n+1)-tuples $(x_1, \ldots, x_n, r)$ of rational numbers (with $r > 0$) . By Proposition 6 this is a countable set of balls.

Let V be any open set in R^n . To show that V is a union of elements of C it suffices to show that for $y \in V$ some $B(x,r) \in C$ contains y and lies in V . (For then V is the union of such $B(x,r)$ -- one for each $y \in V$.) Since V is open some

$$B(y,s) \subset V .$$

Choose x with all coordinates rational such that

$$d(x,y) < \frac{s}{3}$$

and let r be a rational number satisfying $\frac{s}{3} < r < \frac{s}{2}$.

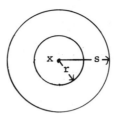

Then $y \in B(x,r)$ and $B(x,r) \subset B(y,s) \subset V$.   q.e.d.

This proposition will be used in an essential way in our study of maximal tori in matrix groups.

E. Manifolds

Definition: By a <u>space</u> we mean some subset of some R^n with the subspace topology. A map

$$f : X \to Y$$

of spaces is a <u>homeomorphism</u> if it is one-to-one, f is continuous, and f^{-1} is continuous.

Example: $f(x) = e^{ix}$ is a one-to-one continuous map of $[0, 2\pi) \subset R$ to the unit circle $S^1 \subset R^2$. But f is not a homeomorphism because f^{-1} is not continuous. (It "tears" the circle open.)

Example: Let G be a matrix group and $x \in G$. Then left translation L_x by x $(L_x(g) = xg)$ is a homeomorphism $L_x : G \to G$. (Exercise.)

A manifold is a space which "locally" looks like some R^n .

Definition: A space X is an n-manifold if each x ∈ X lies in some open set homeomorphic to some B(o,r) ⊂ R^n . An n-manifold is said to have dimension n .

Proposition 8: A matrix group of dimension n is an n-manifold.

Proof: The exponential map from the n-dimensional tangent space T to G is continuous. It is one-to-one on some neighborhorhood V of O in T because it has an inverse (log). Also this inverse is continuous. Take B(o,r) ⊂ V and we have that the identity matrix I has the right kind of neighborhood. For any x ∈ G we have

$$L_x \circ \exp : B(o,r) \to G$$

being a homeomorphism (composition of homeomorphisms is a homeomorphisim) onto a neighborhood of x . Thus G is an n-manifold.

Definition: A manifold is called closed if it is compact (= closed and bounded).

Proposition 9: GL(n,k) is not closed, but ⊙(n,k) is closed.

Proof: Clearly GL(n,k) is not bounded, because for every non-zero real number r , rI ∈ GL(n,k) . (This also shows that GL(n,k) is not closed. Because 0 ∈ M_n(k) - GL(n,k) but every ball with center 0 will contain some rI ∈ GL(n,k) .)

If A ∈ ⊙(n,k) then the rows are unit vectors so that as a vector in M_n(k) the length of A is ≤ n . Thus ⊙(n,k) is a bounded set.

To see that M_n(k) - ⊙(n,k) is open, suppose B ∈ M_n(k) - ⊙(n,k). Then there exists x,y ∈ k^n such that

$$\langle xB, yB \rangle \neq \langle x, y \rangle .$$

Since $\langle \ , \ \rangle$ is continuous, there is some open ball $B(B,s)$ in $M_n(k)$ such that for $B' \in B(B,s)$ we have $\langle xB', yB' \rangle \neq \langle x, y \rangle$. Thus $B' \notin \mho(n,k)$.

q.e.d.

We finish this chapter with a result which will be of substantial use to us later on.

Proposition 10: Let N and M be closed n-manifolds with $N \subset M$. If M is connected, then $N = M$.

Proof: We want to show that $M - N$ is empty. If it isn't, choose $y \in M - N$ and $x \in N$. Since M is connected, there exists a path

$$\rho : [0,1] \to M$$

with $\rho(0) = x$ and $\rho(1) = y$. Then $\rho^{-1}(M - N)$ is an open set in $[0,1]$ (see Exercise 3) and it contains 1 but not 0 . Let t_0 be the largest element of the closed set $I - \rho^{-1}(M - N)$. Then

(i)   every $B(t_0, \epsilon)$ contains points of $\rho^{-1}(M - N)$, but

(ii)   since N is a manifold there is some open neighborhood U of $\rho(t_0)$ in N . By continuity of ρ some $B(t_0, \epsilon)$ maps by ρ into N . This contradiction shows $N = M$.

Corollary: The map $\rho : Sp(1) \to SO(3)$ (see Chapter V) is sur- jective.

Proof: Since ρ is a homeomorphism on some neighborhood of each point, we see that the image $\rho(Sp(1))$ is a 3-manifold. Since ρ is continuous, this image is a closed 3-manifold (Proposition 5). It remains to prove that   SO(3)   is connected (so that we may apply

Proposition 10).   It suffices to show that any $A \in SO(3)$ may be

joined to the identity matrix   I   by a path in   SO(3) .

   We have   det A = 1   and that $\{Ae_1, Ae_2, Ae_3\}$ is an orthonormal

basis for R^3 .   Let   B   be a rotation sending e_1 to Ae_1 and

leaving the direction perpendicular to the (e_1, Ae_1) plane fixed.

(If $Ae_1 = e_1$ proceed directly to the next step.   If e_1 and Ae_1

are antipodal on S^2 ,   then we have two choices for   B .)   Clearly,

there is a path ω from   I   to   B   in   SO(3) .   Now $Be_1 = Ae_1$ so

Be_2 and Be_3 are an orthonormal basis for the plane perpendicular

to Ae_1 .   Let   C   be a rotation of this plane sending Be_2 to Ae_2

and Be_3 to Ae_3 .   (If we can't do this, we would have   det A = -1.)

There is a path σ from   I   to   C   in   SO(3) .   Since   A = BC ,   we

can multiply the paths ω and σ to get a path from   I   to   A   in

SO(3) .   q.e.d.

   Note: In Chapter VIII we will prove that   SO(n)   is connected

for all   n .

F.   Exercises

   1.   Show that the definition of continuity reduces to the usual

$\epsilon - \delta$ definition for $f : (a,b) \to R$.

   2.   Suppose we have $A \subset R^n$ and have functions

$$A \xrightarrow{f} R^m \qquad f(A) \xrightarrow{g} R^p .$$

We have seen that f and g continuous implies that g ∘ f is continuous. Give examples to show that:

  f continuous and g ∘ f continuous $\not\Rightarrow$ g continuous .

  g continuous and g ∘ f continuous $\not\Rightarrow$ f continuous .

3. Show that for $A \subset R^n$ and $f : A \to R^m$ then f is continuous
↔ for each open set U in R^m , $f^{-1}(U)$ is an open set in A .

4. Show that if A,B are connected sets in R^n and $A \cap B \neq \phi$,
then $A \cup B$ is connected.

5. Let H be any connected subgroup of a matrix group G . Show
that

$$S = \bigcup_{x \in G} xHx^{-1}$$

is connected.

6. Show that matrix multiplication is continuous (with one matrix
fixed; i.e., $A \in M_n(k)$, $L_A : M_n(k) \to M_n(k)$ given by $L_A(B) = AB$
is continuous) .

7. Show that an arbitrary union of open sets is an open set.

8. Let $A \subset R^n$ and $x \in R^n$. We say that x is a limit point
of A (x lp A) if every

$$B(x,r) \cap A$$

is an infinite set. Show that $C \subset R^n$ is a closed set ↔
(x lp C $\Rightarrow$ x ∈ C) .

9. Let $D \subset R^n$ be open and closed. Show that if D is not empty, then $D = R^n$. (See the proof of Proposition 10.)

Chapter 7
Maximal Tori

A. Cartesian products of groups

If G and H are groups, we make G × H into a group by defining

$$(g,h)(g',h') = (gg',hh') \ .$$

This works and if G,H are abelian so is G × H . (Exercises.)

Example: If G is a group of n × n matrices and H is a group of m × m matrices, we can represent elements of G × H as (n + m) × (n + m) matrices by

$$(g,h) \ = \ \begin{array}{c} n\{ \\ \\ \end{array} \left(\begin{array}{c|c} g & 0 \\ \hline \underbrace{0}_{n} & \underbrace{h}_{m} \end{array} \right) \begin{array}{c} \\ \\ \}m \end{array} \ .$$

Then matrix multiplication gives the operation described above on G × H . (Exercise.) Let us look at an important special case of this.

Let $G = \{(\begin{smallmatrix} \cos\theta & \sin\theta \\ -\sin\theta & \cos\theta \end{smallmatrix})\}$ and $H = \{(\begin{smallmatrix} \cos\phi & \sin\phi \\ -\sin\phi & \cos\phi \end{smallmatrix})\}$. Then

$$G \times H = \left\{ \begin{pmatrix} \cos\theta & \sin\theta & 0 & 0 \\ -\sin\theta & \cos\theta & 0 & 0 \\ 0 & 0 & \cos\phi & \sin\phi \\ 0 & 0 & -\sin\phi & \cos\phi \end{pmatrix} \right\} \ .$$

Both G and H are circle groups, and the (abelian) group G x H $= S^1 \times S^1$ is called a 2-torus.

Definition: A k-torus is the Cartesian product of k circle groups.

We have seen that a k-torus can be represented by a "block diagonal" 2k x 2k real matrix. But it is easy to see that

$$T = \begin{pmatrix} e^{i\theta_1} & & \\ & e^{i\theta_2} & \bigcirc \\ & \bigcirc & \ddots \\ & & & e^{i\theta_k} \end{pmatrix}$$

is a k-torus, so we can represent a k-torus as diagonal complex k x k matrices.

Proposition 1: If G is an abelian matrix group and γ, σ are one-parameter subgroups, then $\gamma\sigma$ is a one - parameter subgroup.

Proof:
$$
\begin{aligned}
(\gamma\sigma)(s+t) &= \gamma(s+t)\sigma(s+t) \\
&= \gamma(s)\gamma(t)\sigma(s)\sigma(t) \\
&= \gamma(s)\sigma(s)\gamma(t)\sigma(t) \\
&= (\gamma\sigma)(s)(\gamma\sigma)(t) \ .
\end{aligned}
$$

Corollary: If G is an abelian matrix group then $\exp : (TG)_e \to G$ is a homomorphism from the vector group of the tangent space $(TG)_e$ to G at e .

Proof: Let $\xi = \gamma'(0)$, $\eta = \sigma'(0)$ with γ, σ being one-parameter subgroups. Then from Chapter IV we know that $\exp(\xi) = \gamma(1)$ and

$\exp(\eta) = \sigma(1)$. We have that $\gamma\sigma$ is a one-parameter subgroup and $(\gamma\sigma)'(0) = \gamma'(0) + \sigma'(0) = \xi + \eta$ (Chapter III), so $\exp(\xi+\eta)$ $= (\gamma\sigma)(1) = \gamma(1)\sigma(1) = \exp \xi \exp \eta$.

Now we know that exp is 1-1 on some neighborhood V of 0 in $(TG)_e$. So for G abelian we know that $\ker(\exp) = L$ is a discrete subgroup of the vector group $(TG)_e$ (i.e. some neighborhood of 0 contains no point of L except 0). Next we consider when $\exp : (TG)_e \to G$ is surjective.

Proposition 2: Let G be a connected matrix group and let H be any subgroup of G containing an open neighborhood U of e . Then H = G .

Proof: Since $U \subset H$ and H is a subgroup, we must have

$$U^2 = \{xy \mid x,y \in U\}, U^3, U^4, \ldots \text{ all in } H .$$

Thus $$W = U \cup U^2 \cup U^3 \cup \ldots \subset H .$$

Each U^k is open, so W is an open set. (Exercise, Chapter VI.) But W is also closed. For, let x be a limit point of W . Then xU is an open set containing x ($e \in U$) and hence must contain some point of W . Thus

$$xu = u_1 \ldots u_m \text{ for some } u, u_1, u_2, \ldots, u_m \in U .$$

But then $x = u_1 \ldots u_m u^{-1} \in W$. In a connected space G only ϕ and G are both open and closed (Exercise, Chapter VI), $W \neq \phi$ so W = G so H = G .

Corollary: If G is a connected abelian matrix group, then

exp : $(TG)_e \to G$ is a surjective homomorphism with a discrete kernel.

Proof: exp is a homomorphism so $\exp((TG)_e)$ is a subgroup of G and it contains some neighborhood U of e . Thus $\exp(TG)_e = G$.

In the exercises it is proved that this implies that $\exp(TG)_e = T^k \times R^{n-k}$ for some k . So we have

Theorem 1: Any compact connected abelian matrix group G is a torus.

B. Maximal tori in groups

Definition: A subgroup H of a matrix group G is a torus if it is isomorphic with a k-torus for some k. It is a maximal torus if it is not contained in any larger torus subgroup of G .

Proposition 3:

$$T = \left\{ \begin{pmatrix} \cos\theta & \sin\theta & 0 \\ -\sin\theta & \cos\theta & 0 \\ 0 & 0 & 1 \end{pmatrix} \right\}$$

is a maximal torus in SO(3) .

Proof: Clearly T is isomorphic with $\left\{ \begin{pmatrix} \cos\theta & \sin\theta \\ -\sin\theta & \cos\theta \end{pmatrix} \right\}$ which is a circle group and thus is a 1-torus.

Suppose there is a larger torus subgroup T' of SO(3) , i.e.,

$$T \subsetneq T' \subset SO(3) .$$

Since T' would be abelian, we would have: $\phi \in SO(3)$ such that

$\phi \notin T$ but ϕ commutes with every element of T. So it suffices to prove that:

If $\phi \in SO(3)$ commutes with each $t \in T$, then $\phi \in T$. Refer back to our proof that Center $SO(3) = \{I\}$ in Chapter V, and you will see that we have already proved this fact.

Proposition 4:

$$T = \left\{ \begin{pmatrix} \cos\theta_1 & \sin\theta_1 & 0 & 0 \\ -\sin\theta_1 & \cos\theta_1 & 0 & 0 \\ 0 & 0 & \cos\theta_2 & \sin\theta_2 \\ 0 & 0 & -\sin\theta_2 & \cos\theta_2 \end{pmatrix} \right\}$$

is a maximal torus in $SO(4)$.

Proof: This clearly is a 2-torus and is a subgroup of $SO(4)$. As before, it suffices to prove that if $\phi \in SO(4)$ commutes with all elements of T then $\phi \in T$.

Let V be the 2-plane in R^4 spanned by e_1, e_2 and W be the 2-plane in R^4 spanned by e_3, e_4. We see that T consists of all pairs

$$(\text{rotation of } V, \text{ rotation of } W).$$

Claim: $\qquad\qquad\qquad\qquad \phi(e_1) \in V$.

Take $\alpha \in T$ such that α is the identity on V but is not the identity on W. Then

$$\phi(e_1) = ae_1 + be_2 + ce_3 + de_4$$
$$\alpha\phi(e_1) = ae_1 + be_2 + c'e_3 + d'e_4$$
$$\phi\alpha(e_1) = \phi(e_1) = ae_1 + be_2 + ce_3 + de_4 \ .$$

This shows $c = 0 = d$, so $\phi(e_1) \in V$. The same kind of proof shows $\phi(e_2) \in V$. Dually, $\phi(e_3)$ and $\phi(e_4)$ are in W .

So we know that ϕ is <u>orthogonal on</u> V and is <u>orthogonal on</u> W . A priori, it could be a <u>reflection</u> in each and we would still have $\phi \subset SO(4)$. But ϕ commutes with all rotations on V and is thus a rotation on V . (See the proof of Proposition 3, Chapter V.) Similarly, ϕ is a rotation on W , so $\phi \in T$. q.e.d.

From Propositions 1 and 2 the general result about maximal tori in SO(n) should be clear. We have n/2 of the 2 × 2 rotation matrices for n even and have a 1 in the n,n position for n odd. The proof of the general case is an obvious extension of the above proofs.

<u>Proposition</u> 5:

$$T = \left\{ \begin{pmatrix} e^{i\theta_1} & & & \bigcirc \\ & e^{i\theta_2} & & \\ & & \ddots & \\ \bigcirc & & & e^{i\theta_n} \end{pmatrix} \right\}$$

<u>is a maximal torus in</u> U(n) .

<u>Proof</u>: Let $\phi \in U(n)$ commute with each $\alpha \in T$. Consider any α of the form

$$\alpha = \begin{pmatrix} 1 & & & \bigcirc \\ & e^{i\theta_2} & & \\ & & \ddots & \\ \bigcirc & & & e^{i\theta_n} \end{pmatrix} \in T .$$

Then $\alpha\phi(e_1) = \phi(\alpha(e_1)) = \phi(e_1)$. So α leaves $\phi(e_1)$ fixed, but such α's can move any vector which is not a multiple of e_1 . Hence

$$\phi(e_1) = \lambda_1 e_1 , \quad \lambda_1 \in \mathbb{C} .$$

Similar arguments give

$$\phi(e_j) = \lambda_j e_j \quad \text{for} \quad j = 1,\ldots,n .$$

Thus

$$\phi = \begin{pmatrix} \lambda_1 & & & \bigcirc \\ & \lambda_2 & & \\ & & \ddots & \\ \bigcirc & & & \lambda_n \end{pmatrix}$$

and, since $\phi \in U(n)$, each λ_j is of unit length. Thus $\phi \in T$ and T is maximal.

Proposition 6:

$$T = \left\{ \begin{pmatrix} e^{i\theta_1} & & \bigcirc \\ & \ddots & \\ \bigcirc & & e^{i\theta_n} \end{pmatrix} \middle| \quad \theta_1 + \ldots + \theta_n = 0 \right\}$$

is a maximal torus in $SU(n)$.

Proof: A matrix α of the form $\begin{pmatrix} e^{i\theta_1} & \bigcirc \\ & \ddots \\ \bigcirc & e^{i\theta_n} \end{pmatrix}$ has

$\det \alpha = e^{i\theta_1} \ldots e^{i\theta_n} = e^{i(\theta_1+\ldots+\theta_n)}$ so that $\alpha \in SU(n) \Leftrightarrow \det \alpha$
$= 1 \Leftrightarrow \theta_1 + \ldots + \theta_n = 0$. So the T described here is just the intersection of $SU(n)$ with the maximal torus given for $U(n)$.

First we must check that this is an (n-1)-torus. To do this, use

$$
\begin{pmatrix} e^{i\theta_1} & & \bigcirc \\ & \ddots & \\ \bigcirc & & e^{i\theta_n} \end{pmatrix} \leftrightarrow \begin{pmatrix} e^{i(\theta_1-\theta_n)} & & & \bigcirc \\ & e^{i(\theta_2-\theta_n)} & & \\ & & \ddots & \\ \bigcirc & & e^{i(\theta_{n-1}-\theta_n)} & \\ & & & 1 \end{pmatrix}
$$

$$\Sigma\theta_i = 0$$

It is an exercise to show that this works.

Now for $n > 2$ the same proof as used for $U(n)$ will work, but for $n = 2$

$$
T = \begin{pmatrix} e^{i\theta} & 0 \\ 0 & e^{-i\theta} \end{pmatrix}
$$

and we do not have matrices $\alpha = \begin{pmatrix} 1 & 0 \\ 0 & e^{-i\theta} \end{pmatrix}$ to use. But for $n = 2$

we give a direct simple proof. If

$$\phi = \begin{pmatrix} a & b \\ c & d \end{pmatrix} \in SU(2) \quad \text{and} \quad \alpha = \begin{pmatrix} i & 0 \\ 0 & -i \end{pmatrix} \in T$$

then

$$\phi\alpha = \begin{pmatrix} ai & -bi \\ ci & -di \end{pmatrix} = \begin{pmatrix} ai & bi \\ -ci & -di \end{pmatrix} = \alpha\phi \quad .$$

Thus $b = 0 = c$ and $\phi \in T$.

Proposition 7: The maximal torus given for $U(n)$ is also a maximal torus for $Sp(n)$.

Proof: Just as for $U(n)$ we can show that any element of $Sp(n)$

which commutes with all $\begin{pmatrix} e^{i\theta_1} & & \bigcirc \\ & \ddots & \\ \bigcirc & & e^{i\theta_n} \end{pmatrix}$ must be diagonal with the

diagonal elements having length 1 . But now these elements are quaternions. However (Exercise), any quaternion which commutes with i must be a complex number.

q.e.d.

C. Centers again

Now that we know maximal tori in our matrix groups, we are able to calculate the centers.

Proposition 8: Center $(Sp(n)) = \{I, -I\}$.

Proof: We have seen that if any element commutes with all elements of the maximal torus we have described, then it must lie in that maximal torus. Hence, in every case

$$\text{Center} \subset T .$$

If $A \in$ Center $Sp(n)$, then $A = \begin{pmatrix} e^{i\theta_1} & & \bigcirc \\ & \ddots & \\ \bigcirc & & e^{i\theta_n} \end{pmatrix}$. Since A must

commute with the matrix jI , it follows that the diagonal elements must be real (and since they are of unit length) they are ± 1 . It is an exercise to show that they all have the same sign. Now I and -I are in the center.

q.e.d.

Proposition 9:   Center $U(n) = \{e^{i\theta}I\} \cong S^1$

Center $SU(n) = \{wI \mid w^n = 1\}$.

Proof: If $B \in$ Center $U(n)$ we get that B is diagonal with diagonal elements complex numbers of unit length.   Let

$$B = \begin{pmatrix} \alpha_1 & & & \\ & \alpha_2 & & \\ & & \ddots & \\ & & & \alpha_n \end{pmatrix} \quad \text{with each} \quad |\alpha_\ell| = 1 . \quad \text{Let}$$

$$A = \begin{pmatrix} 0 & 1 & 0 & \dots & 0 \\ 1 & 0 & 0 & \dots & 0 \\ 0 & 0 & 1 & & \\ \vdots & & & \ddots & 0 \\ \vdots & & 0 & & 1 \end{pmatrix} .$$

Then $AB = BA$ shows $\alpha_1 = \alpha_2$, etc., so all α are equal. Clearly any $e^{i\theta}I$ is in the center, so the center of $U(n)$ is as asserted.

For $SU(n)$ we note that the same argument will show that an element must be of the form $e^{i\theta}I$ to be in the center. But

$$\det(e^{i\theta}I) = e^{in\theta} ,$$

and since this must be 1 , $e^{i\theta}$ must be an nth root of unity.

q.e.d.

So Center $SU(n) = SU(n) \cap$ Center $U(n)$.

Finally, we want to calculate the center of $SO(n)$. It turns out that it depends on whether n is even or odd. The groups $SO(2n)$ and the groups $SO(2n+1)$ are different in some important ways.

Now $SO(2) = S^1$ is abelian so its center is the group itself. We have already proved that the center of $SO(3)$ is just $\{I\}$.

<u>Claim</u>: For $k \geq 3$ any element in the center of $SO(k)$ must be a diagonal matrix.

As before, if $A \in SO(k)$ is in the center, it must be in our standard maximal torus. So suppose $A \in$ Center $SO(k)$ is of the form

$$A = \begin{pmatrix} \cos \theta_1 & \sin \theta_1 & 0 \ldots \\ -\sin \theta_1 & \cos \theta_1 & 0 \ldots \\ 0 & 0 & * \ldots \end{pmatrix}$$

Let

$$P = \begin{pmatrix} 0 & 0 & 1 & & \\ 0 & 1 & 0 & \bigcirc & \\ -1 & 0 & 0 & & \\ & & & 1 \cdot_{\cdot} 0 & \\ & \bigcirc & & 0 \cdot_{\cdot} 1 \end{pmatrix} \in SO(k) \quad .$$

Then PA has zero in the $1,2$ position, whereas AP has $\sin \theta_1$ in the $1,2$ position. Thus $\sin \theta_1 = 0$. Similar arguments show all off diagonal terms are zero.

It follows also (since each $\sin \theta_i = 0$) that each diagonal term $(\cos \theta_i)$ is 1 or -1. So each 2×2 block is $\begin{pmatrix} 1 & 0 \\ 0 & 1 \end{pmatrix}$ or $\begin{pmatrix} -1 & 0 \\ 0 & -1 \end{pmatrix}$. Arguments like we used for $U(n)$ show that all diagonal terms must be equal. So we finally conclude that

$$\text{Center } SO(2n+1) = \{I\}$$

$$\text{Center } SO(2n) = \{I, -I\} \quad .$$

For example, $\begin{pmatrix} -1 & & & \bigcirc \\ & -1 & & \\ & & -1 & \\ \bigcirc & & & -1 \end{pmatrix}$ is in the center of $SO(4)$, but $\begin{pmatrix} -1 & & \bigcirc \\ & -1 & \\ \bigcirc & & -1 \end{pmatrix}$

is not even in $SO(3)$.

We now tabulate the information we have generated about our groups.

Dimensions, Centers, Maximal tori

Group	Dimension	Center	Standard Maximal torus
U(n)	n^2	$\{e^{i\theta}I\} \cong S^1$	$\left\{\begin{pmatrix} e^{i\theta_1} & & \bigcirc \\ & \ddots & \\ \bigcirc & & e^{i\theta_n} \end{pmatrix}\right\}$
SU(n)	n^2-1	$\{wI \mid w^n=1\} \cong \frac{\mathbb{Z}}{n}$	$\left\{\begin{pmatrix} e^{i\theta_1} & & \bigcirc \\ & \ddots & \\ \bigcirc & & e^{i\theta_n} \end{pmatrix} \middle| \Sigma\theta_i = 0\right\}$
SO(2n+1)	$\frac{(2n+1)(2n)}{2} = 2n^2+n$	$\{I\}$	$\left\{\begin{pmatrix} \text{rot } \theta_1 & & \bigcirc \\ & \ddots & \\ \bigcirc & & \text{rot } \theta_{n_1} \end{pmatrix}\right\}$
SO(2n)	$\frac{2n(2n-1)}{2} = 2n^2-n$	$\{I,-I\}$	$\left\{\begin{pmatrix} \text{rot } \theta_1 & & \bigcirc \\ & \ddots & \\ \bigcirc & & \text{rot } \theta_n \end{pmatrix}\right\}$
Sp(n)	$2n^2+n$	$\{I,-I\}$	$\left\{\begin{pmatrix} e^{i\theta_1} & & \bigcirc \\ & \ddots & \\ \bigcirc & & e^{i\theta_n} \end{pmatrix}\right\}$

Note that we have nothing to distinguish $SO(2n+1)$ and $\frac{Sp(n)}{Center}$. A good part of the remainder of this book is devoted to deciding for which n these are isomorphic.

D. Exercises

1. Show that the operation defined on $G \times H$ does make it into a group. Prove that if G and H are abelian, so is $G \times H$.

2. Do the exercise in the first example of §A.

3. Show directly that the product of the matrices

$$\begin{pmatrix} \cos \theta & \sin \theta \\ -\sin \theta & \cos \theta \end{pmatrix} \text{ and } \begin{pmatrix} \cos \phi & \sin \phi \\ -\sin \phi & \cos \phi \end{pmatrix}$$

is the matrix for a rotation through angle $\theta + \phi$.

4. Let T be a maximal torus in a matrix group G and let $x \in G$. Prove that xTx^{-1} is also a maximal torus in G .

5. Prove that if q is a quaternion such that $qi = iq$, then q is a complex number.

6. Show that $\dfrac{U(n)}{\text{center}} \cong \dfrac{SU(n)}{\text{center}}$.

7. A lattice subgroup K of R^n consists of all integar linear combinations of some set of linearly independent vectors. More explicitely, let

$$v_1, v_2, \ldots, v_k$$

be linearly independent vectors in R^n . For any integers $a_1, \ldots, a_k$ we form

$$a_1 v_1 + \ldots + a_k v_k .$$

Then $K = \{a_1 v_1 + \ldots + a_k v_k\} \mid a_i \in Z\}$. It is routine to verify that K is a subgroup of R^n .

A subgroup H of R^n is <u>discrete</u> if some neighborhood of 0 in R^n contains no point of H other than 0 . Prove that: A discrete subgroup of R^n is a lattice subgroup. (Choose a nonzero vector $v_1 \in H$ such that no element of H lies in the interval $[0, v_1]$ in Rv_1 . Show that Rv_1 contains all integral multiples of v_1 but no other elements of H . Choose v_2 in H with v_2 not an integral multiple of v_1 . Show that the span of v_1 and v_2 in R^n contains all integral linear combinations of v_1 and v_2 but no other elements of H . etc.)

8. Show that if L is a lattice group in R^n generated by $v_1, \ldots, v_k$. Then R^n/L is isomorphic with the product of k-torus and R^{n-k} .

Chapter 8
Covering by Maximal Tori

A. General remarks

In Exercise 4 of Chapter VII one showed that if T is a maximal torus in a matrix group G , then for any $x \in G$, xTx^{-1} is also a maximal torus. What we prove in this chapter is that if T is our standard maximal torus in one of our connected matrix groups G , then

$$(+) \qquad\qquad G = \bigcup_{x \in G} xTx^{-1}$$

showing that every element of G lies in at least one maximal torus.

To say that $G = \bigcup_{x \in G} xTx^{-1}$ is to say that given $y \in G$ there exists $x \in G$ such that

$$y \in xTx^{-1} .$$

This is equivalent to: Given $y \in G$, there exists $z \in G$ such that $zyz^{-1} \in T$ (take $x = z^{-1}$) . So we want to show that $y \in G$ can be put in diagonal or 2×2 block-diagonal form by conjugation.

We begin by reviewing a little linear algebra. Let V be a vector space over a field k and let

$$\phi : V \to V$$

be a linear map.

Definition: A subspace W of V is ϕ - stable if $\phi(W) \subset W$. In this case, we can restrict the domain of ϕ to W to get a linear map

$$\phi | W : W \to W \quad .$$

Examples: Let $V = R^3$ and W be the 2-plane spanned by e_1 and e_2. If

$$\phi = \begin{pmatrix} \cos \theta & \sin \theta & 0 \\ -\sin \theta & \cos \theta & 0 \\ 0 & 0 & 1 \end{pmatrix}$$

then W is ϕ - stable.

Let $V = R^2$ and $\phi = \begin{pmatrix} 0 & 1 \\ 1 & 0 \end{pmatrix}$. Then $W = \{(x,x)\}$ is ϕ - stable and so is $W' = \{(x,-x)\}$.

An important special case occurs when dim W = 1.

Definition: A nonzero vector $v \in V$ is an eigenvector for ϕ if there exists a $\lambda \in k$ such that

$$\phi(v) = \lambda v \quad .$$

λ is the corresponding eigenvalue.

Now if ϕ has one eigenvector v then it has a "line" of them; namely, if $r \in k$ with $r \neq 0$, then

$$\phi(rv) = r\phi(v) = r\lambda v = \lambda(rv) \quad .$$

(We will have to be a little more careful when using our "skew" field $\mathbb{H}$.) Thus rv is also an eigenvector with the same eigenvalue λ. This suggests that eigenvalues may be more fundamental.

<u>Definition</u>: For any $\lambda \in k$, set

$$V(\lambda) = \{v \in V \mid \phi(v) = \lambda v\} \quad .$$

Thus $V(\lambda)$ includes the zero vector o and all eigenvectors having eigenvalue λ . It is easy to see that $V(\lambda)$ is a subspace of V , the <u>eigenspace</u> belonging to λ .

For example, taking $\lambda = 0$, we see that

$$V(0) = \{v \in V \mid \phi(v) = 0\}$$

is just the kernel (or null space) of the linear map ϕ .

Now take $\dim V$ to be finite. We may as well take $V = k^n$. Then linear maps $\phi : V \to V$ correspond 1-1 with elements of $M_n(k)$, so we can think of $\phi \in M_n(k)$.

<u>Proposition</u> 1: $V(\lambda) \neq \{o\} \Leftrightarrow \det(\phi - \lambda I) = 0$.

<u>Proof</u>: $\Rightarrow$

If $o \neq v$ satisfies $\phi(v) = \lambda v$, then $(\phi - \lambda I)v = o$ showing that $\phi - \lambda I$ is singular. Thus $\det(\phi - \lambda I) = 0$.

$\Leftarrow$

If $\det(\phi - \lambda I) = 0$, then $\phi - \lambda I$ must send some nonzero vector v to o ; i.e., $\phi(v) = \lambda v$, so $V(\lambda) \neq \{o\}$.

B. (+) <u>for</u> $U(n)$ <u>and</u> $SU(n)$

We are concerned with $\mathbb{c}$ - linear maps $\mathbb{c}^n \to \mathbb{c}^n$ which we represent by elements of $M_n(\mathbb{C})$. We begin with two easy observations:

(a) If A is unitary and $W \subset \mathbb{C}^n$ is A - stable, then $A|W$ is unitary.

If $x,y \in W$ then xA and $yA \in W$ and

$$\langle xA, yA \rangle = \langle x, y \rangle .$$

(b) If A is unitary and $W \subset \mathbb{C}^n$ is A - stable, then the orthogonal complement

$$W^{\perp} = \{x \in \mathbb{C}^n \mid \langle x, y \rangle = 0 \text{ for all } y \in W\}$$

is also A - stable.

Let $x \in W^{\perp}$. For any $y \in W$ we have

$$\langle xA, y \rangle = \langle x, y^t \overline{A} \rangle = \langle x, yA^{-1} \rangle .$$

Now A is an isomorphism of W onto W so $yA^{-1} \in W$. Thus

$$\langle xA, y \rangle = 0 \text{ for all } y \in W , \text{ showing } xA \in W^{\perp} .$$

Proposition 2: For $A \in U(n)$ there exists an orthonormal basis of eigenvectors $v_1, \ldots, v_n \in \mathbb{C}^n$ for A .

Proof: By (a) and (b) above, it suffices to show that a unitary matrix C always has an eigenvector v . For we can make v unit length and restrict C to $v^{\perp}$, etc.

But let $A \in M_n(\mathbb{C})$. Then

$$p(\lambda) = \det(A - \lambda I)$$

is a polynomial in λ and $\mathbb{C}$ is algebraically closed so it has a root. Thus A has an eigenvector.

Proposition 3: The conjugates of $T = \begin{pmatrix} e^{i\theta_1} & & \bigcirc \\ & \ddots & \\ \bigcirc & & e^{i\theta_n} \end{pmatrix}$

cover U(n) .

Proof: Given $A \in U(n)$ we will find $B \in U(n)$ such that

$$BAB^{-1} \in T \ .$$

Let $v_1, \ldots, v_n$ be an orthonormal basis consisting of eigenvectors of A . Let B send

$$e_j \quad \text{to} \quad v_j \ .$$

Then BAB^{-1} sends e_j to v_j to $\lambda_j v_j$ to $\lambda_j e_j$. So

$$BAB^{-1} = \begin{pmatrix} \lambda_1 & & & \bigcirc \\ & \lambda_2 & & \\ & & \ddots & \\ \bigcirc & & & \lambda_n \end{pmatrix} \ .$$

Since BAB^{-1} is unitary, each $\lambda_j e_j$ is a unit vector; i.e., $|\lambda_j| = 1$. Thus $BAB^{-1} \in T$.   q.e.d.

Corollary: The conjugates of

$$T = \left\{ \begin{pmatrix} e^{i\theta_1} & & \bigcirc \\ & \ddots & \\ \bigcirc & & e^{i\theta_n} \end{pmatrix} \ \middle| \ \Sigma\theta_j = 0 \right\}$$

cover SU(n) .

Proof: Given $A \in SU(n)$ we want $B \in SU(n)$ such that $BAB^{-1} \in T$. If we take $B' \in U(n)$ such that

$$B'A(B')^{-1} \text{ is in the maximal torus for } U(n) ,$$

it is actually in the maximal torus for $SU(n)$ since

$$\det(B'A(B')^{-1}) = \det A = 1 .$$

If we choose $u \in C$ such that $\det B' = \dfrac{1}{u^n}$ and set

$$B = uB' ,$$

then $BAB^{-1} = B'A(B')^{-1}$ and $B \in SU(n)$.

C. (+) <u>for</u> SO(n)

For any matrix group G and maximal torus T we have that

$$\bigcup_{x \in G} xTx^{-1}$$

is a connected set, so if G is not connected, (+) could not hold. Since $\Theta(n)$ is not connected, we know that

$$\Theta(n) \neq \bigcup_{x \in \Theta(n)} xTx^{-1} .$$

But we will prove (+) for $SO(n)$ and this will, incidentally, prove that

$$SO(n) \text{ is connected } .$$

Clearly we can have no result like Proposition 2 for SO(n) .
For example $A = \begin{pmatrix} \cos \theta & \sin \theta \\ -\sin \theta & \cos \theta \end{pmatrix} \in SO(2)$ with $A \neq I$ has no nonzero
eigenvector. But this is about all that can happen--we can find
stable 2-planes. To see this we consider symmetric linear maps.

Definition: A linear map $S : R^n \to R^n$ is symmetric if

$$\langle xS, y \rangle = \langle x, yS \rangle \quad \text{for all} \quad x, y \in R^n \; .$$

Note that for $A \in M_n(R)$, $S = A + {}^tA$ is symmetric. For we
have

$$\langle xS, y \rangle = \langle xA, y \rangle + \langle x{}^tA, y \rangle = \langle x, y{}^tA \rangle + \langle x, yA \rangle = \langle x, Sy \rangle \; .$$

Corresponding to (a) and (b) for U(n) we have: If S is
symmetric and $W \in R^n$ is S-stable, then $S|W$ is symmetric and $W^\perp$
is S-stable. Thus, just as for U(n) , it suffices to show that a
symmetric matrix always has an eigenvector.

Proposition 4: A (real) symmetric matrix S has an eigenvector.

Proof: Define $\phi : R^n - (0) \to R$ by

$$\phi(x) = \frac{\langle x, xS \rangle}{\langle x, x \rangle} \; .$$

Note that for nonzero $r \in R$ we have $\phi(rx) = \phi(x)$. Thus if S^{n-1}
is the unit sphere, we have

$$\phi(R^n - (0)) = \phi(S^{n-1}) \; .$$

Now ϕ is continuous and S^{n-1} is compact. Thus $\phi(S^{n-1})$ is compact and we can find $v \in S^{n-1}$ such that

$\Phi(v)$ is an <u>absolute</u> <u>maximum</u> .

We claim that v is then an eigenvector for S .

For any nonzero $y \in R^n$, if we set

$$f(t) = \Phi(v + ty) ,$$

we must have $f'(0) = 0$. This will allow us to show v is an eigenvector and to find its eigenvalue.

We have

$$f(t) = \frac{\langle v+ty, vS+tyS \rangle}{\langle v+ty, v+ty \rangle} = \frac{\alpha(t)}{\beta(t)} ,$$

$$f'(0) = \frac{\alpha'(0)\beta(0) - \beta'(0)\alpha(0)}{(\beta(0))^2} .$$

Since v is of unit length, $\beta(0) = 1$ so that

$$f'(0) = \alpha'(0) - \beta'(0)\alpha(0) .$$

We easily calculate that $\beta'(0) = 2\langle v,y \rangle$ and $\alpha(0) = \langle v,vS \rangle$ and $\alpha'(0) = \langle y,vS \rangle + \langle v,yS \rangle = 2\langle vS,y \rangle$ (S and $\langle \, , \, \rangle$ are symmetric). This gives $f'(0) = 2\langle vS,y \rangle - 2\langle v,vS \rangle\langle v,y \rangle = 0$, or

$$\langle (vS - \langle v,vS \rangle v), y \rangle = 0 .$$

Now $\langle \, , \, \rangle$ is nondegenerate and y is arbitrary, so

$$vS = \langle v,vS \rangle v . \qquad\qquad \text{q.e.d.}$$

We first use Proposition 4 to study $\mathfrak{G}(n)$ and then prove (+) for $SO(n)$.

Proposition 5: Let $A \in \mathbb{O}(n)$; then there exists an A-stable subspace W of $\mathbb{R}^n$ with

$$\dim W \in \{1,2\} \ .$$

Proof: $S = A + {}^tA$ is symmetric. Let w be an eigenvector for S. Consider w and wA.

case (i) If w and wA are linearly dependent, $wA = \lambda w$, then $W = \{rw \mid r \in \mathbb{R}\}$ is A-stable and $\dim W = 1$.

case (ii) If linearly independent, let $W = \text{Span}(w, wA)$. We have $wS = \lambda w$, i.e., $wA + w{}^tA = \lambda w$ and ${}^tA = A^{-1}$ so that

$$wA + wA^{-1} = \lambda w \quad \text{or} \quad wA^2 + w = \lambda wA \ .$$

This implies that W is A-stable. For, let

$$\alpha w + \beta wA \in W \ .$$

Then

$$(\alpha w + \beta wA)A = \alpha wA + \beta wA^2$$

$$= \alpha wA + \beta(\lambda wA - w)$$

$$= (-\beta)w + (\alpha+\beta\lambda)wA \in W \ .$$

Proposition 6: The conjugates of our standard maximal torus in $SO(n)$ cover $SO(n)$.

Proof: We choose an orthonormal basis $v_1, \dots, v_n$ using $A \in SO(n)$ as follows.

If A has a stable one-dimensional subspace, choose one and take

v_1 to be a unit vector in it. Next do the same for $v_1^\perp$, and continue as long as possible, generating an orthonormal set $v_1, \ldots, v_k$ of eigenvectors for A . Let $W = \text{span}(v_1, \ldots, v_k)^\perp$. A is orthogonal on W . Choose a stable 2-space W_1 and let v_{k+1}, v_{k+2} be an orthonormal basis for W_1 , etc. This gives an orthonormal basis

$$v_1, v_2, \ldots, \underbrace{v_k, v_{k+1}, v_{k+2}}_{W_1}, \ldots, \underbrace{v_{n-1}, v_n}_{\frac{W_{n-k}}{2}} .$$

We use this to find $B \in SO(n)$ such that

$$BAB^{-1} \in T = \text{our standard maximal torus.}$$

Let C map e_i to v_i . Then $C \in \mathcal{O}(n)$ and

$$CAC^{-1} \in T .$$

(Note that $\det CAC^{-1} = \det A = 1$.) If it happens that $\det C = 1$, let $B = C$. If $\det C = -1$ we seek $D \in \mathcal{O}(n) - SO(n)$ such that $DTD^{-1} = T$. Then $B = DC$ will satisfy $BAB^{-1} \in T$ and $\det B = 1$. Such a D is easy to find; e.g.,

$$D = \begin{pmatrix} 0 & 1 & & & \\ 1 & 0 & & \bigcirc & \\ & & 1 & & \\ & \bigcirc & & \ddots & \\ & & & & 1 \end{pmatrix} .$$

This proves (+) for $SO(n)$.

D. (+) <u>for</u> Sp(n)

Our proof that conjugates of $T = \begin{pmatrix} e^{i\theta_1} & & \bigcirc \\ & \ddots & \\ \bigcirc & & e^{i\theta_n} \end{pmatrix}$ cover Sp(n)

will relyy on the following property of $\mathbb{H}$.

<u>Proposition</u> 7: $\mathbb{H}$ <u>has square roots</u>.

<u>Proof</u>: Given $q \in \mathbb{H}$ we must find $h \in \mathbb{H}$ such that $h^2 = q$.
We may as well take $|q| = 1$ (if $q = 0$ take $h = 0$) and may then
assume $|h| = 1$. If $q = x + iy + jz + kd$ set $h = \alpha + i\beta + j\gamma + k\delta$
$h^2 = q$ gives

$$\alpha^2 - \beta^2 - \gamma^2 - \delta^2 = x$$

$$2\alpha\beta = y \qquad 2\alpha\gamma = z \qquad 2\alpha\delta = w$$

Since $|q| = 1 = |h|$, these give $2\alpha^2 = 1 + x$. Choose α satis-
fying this and then solve for β, γ, δ .

Next we find stable subspaces for $A \in Sp(n)$ much as we did
for orthogonal matrices.

<u>Proposition</u> 8: <u>For</u> $A \in Sp(n)$ <u>there exists a nonzero</u> $v \in \mathbb{H}^n$
<u>such that</u>

$$W = \text{Span}(v, vA) \quad \underline{is} \ A - \underline{stable}.$$

<u>Proof</u>: Define $\Phi : \mathbb{H}^n - (o) \to \mathbb{H}$ by

$$\Phi(x) = \frac{\langle xA, x \rangle}{\langle x, x \rangle} \ .$$

For any nonzero real number r we have $\Phi(rx) = \Phi(x)$ so that Φ is constant on real rays through o in $\mathbb{H}^n - (o)$. So we can find

$$v \in S^{4n-1}$$

making Φ an absolute maximum. For arbitrary $y \in \mathbb{H}^n$ we set

$$f(t) = \Phi(v + ty) \ , \quad t \in \mathbb{R} \ ,$$

and we have $f'(0) = 0$. This gives

$$0 = \langle yA, v \rangle + \langle vA, y \rangle - (\langle y, v \rangle + \langle v, y \rangle)\langle vA, v \rangle$$

or

$$0 = \langle (vA - \langle vA, v \rangle)v, y \rangle + \langle y, (vA^{-1} - \langle vA, v \rangle v) \rangle \ .$$

Choose y to be an arbitrary but <u>real</u> n-tuple and we see that

$$\langle vA + vA^{-1} - 2\langle vA, v \rangle v, y \rangle = 0$$

for all real n-tuples y. This implies that'

$$vA + vA^{-1} = 2\langle vA, v \rangle v$$

or

$$(*) \qquad\qquad vA^2 + v = 2\langle vA, v \rangle vA \ .$$

Just as in Proposition 5, such a formula proves $W = \text{Span}(v, vA)$ is A - stable.

<u>Proposition</u> 9: <u>Any</u> $A \in Sp(n)$ <u>has</u> <u>an</u> <u>eigenvector.</u>

<u>Proof</u>: Choose v such that $\text{Span}(v, vA)$ is A - stable. We

claim that A has an eigenvector of the form

$$w = \alpha v + \beta (vA) \quad , \quad \alpha, \beta \in \mathbb{H} \quad .$$

We have

$$wA = \alpha vA + \beta (vA^2)$$

$$= \alpha vA + \beta (2\langle vA, v \rangle vA - v) \qquad \text{from } (*)$$

$$= (\alpha + 2\beta \langle vA, v \rangle) vA - \beta v \quad .$$

We seek $\lambda \in \mathbb{H}$ such that $wA = \lambda w$, or

$$(\alpha + 2\beta \langle vA, v \rangle) vA - \beta v = \lambda \alpha v + \lambda \beta (vA) \quad .$$

So we want $\beta = -\lambda \alpha$

$$(\alpha + 2\beta \langle vA, v \rangle)(vA) = \lambda \beta (vA) \quad .$$

This gives (putting $-\lambda \alpha$ in for β)

$$\alpha (\lambda^2 - 2\langle vA, v \rangle + 1) = 0 \quad .$$

Since $\mathbb{H}$ has square roots, we can find λ such that
$\lambda^2 - 2\langle vA, v \rangle + 1 = 0$.   q.e.d.

Proposition 10: $(\ast)$ holds for Sp(n) .

Proof: Just as for U(n) we now find an orthonormal basis of eigenvectors and thus $B \in $ Sp(n)   such that

$$BAB^{-1} = \begin{pmatrix} \lambda_1 & & \bigcirc \\ & \ddots & \\ \bigcirc & & \lambda_n \end{pmatrix}$$

with the $\lambda_1, \ldots, \lambda_n$ quarternions of unit length. If we conjugate this by a diagonal matrix

$$Q = \begin{pmatrix} q_1 & & \bigcirc \\ & \ddots & \\ \bigcirc & & q_n \end{pmatrix} \in Sp(n) , \quad |q_j| = 1 ,$$

we get

$$QBA(QB)^{-1} .$$

This will be in the standard maximal torus if each $q_j \lambda_j q_j^{-1} = q_j \lambda_j \bar{q}_j$ is a complex number. It is proved in Exercise 5 that we can choose the q_j's to do this.

E. **Relfections in** R^n **(again)**

Let $x \in R^n$ be nonzero and let $x^\perp$ be the $(n-1)$-hyperplane through o perpendicular to x .

Proposition 11: The reflection $\phi : R^n \to R^n$ in $x^\perp$ is given by the formula

$$\phi(y) = y - 2 \frac{\langle x, y \rangle}{\langle x, x \rangle} x .$$

Proof:

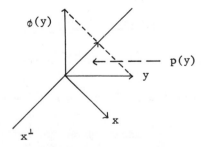

Let p(y) be the projection of y in $x^{\perp}$. Then y - p(y) is some multiple of x , say rx , and $\langle p(y),x \rangle = 0$. This gives $r = \dfrac{\langle x,y \rangle}{\langle x,x \rangle}$. So

$$p(y) = y - \frac{\langle x,y \rangle}{\langle x,x \rangle} x \quad \text{and} \quad \phi(y) - 2\frac{\langle x,y \rangle}{\langle x,x \rangle} x \quad .$$

We can check directly that ϕ is an orthogonal map.

$$\langle \phi(y),\phi(z) \rangle = \langle y-2\frac{\langle x,y \rangle}{\langle x,x \rangle} x, \ z-2\frac{\langle x,z \rangle}{\langle x,x \rangle} x \rangle$$

$$= \langle y,z \rangle -2\frac{\langle x,z \rangle}{\langle x,x \rangle}\langle y,x \rangle -2\frac{\langle x,y \rangle}{\langle x,x \rangle}\langle x,z \rangle +4\frac{\langle x,y \rangle\langle x,z \rangle}{\langle x,x \rangle\langle x,x \rangle}\langle x,x \rangle$$

and this is just $\langle y,z \rangle$.

Also it is easy to see that det ϕ = -1 because we can choose a basis with x as first vector and the other vectors forming a basis for $x^{\perp}$. Relative to this basis ϕ is given by

$$\begin{pmatrix} -1 & & \bigcirc \\ & 1 & \\ \bigcirc & & 1 \end{pmatrix} \quad .$$

Proposition 12: If $A \in \mathcal{O}(n)$ is reflection in a hyperplane W in $\mathbb{R}^n$ and $B \in \mathcal{O}(n)$, then

$$BAB^{-1}$$

is reflection in the hyperplane WB^{-1} .

Proof: Let $w' = wB^{-1} \in WB^{-1}$. Then

$$w'BAB^{-1} = wB^{-1}BAB^{-1} = wAB^{-1} = wB^{-1} = w' \ ,$$

so that BAB^{-1} is the identity on WB^{-1} .

Let $\alpha_1, \ldots, \alpha_n$ be an orthonormal basis with α_1 perpendicular to W and $\alpha_2, \ldots, \alpha_n$ a basis fpr W . Then BAB^{-1} is the identity on $\alpha_2 B^{-1}, \ldots, \alpha_n B^{-1}$ and

$$(\alpha_1 B)(B^{-1}AB) = \alpha_1 AB = -\alpha_1 B ,$$

proving the proposition.

Proposition 13: $\mathfrak{S}(n)$ is generated by reflections.

Proof: We want to show that any element of $\mathfrak{S}(n)$ may be written as the product of a finite number of reflections.

First we prove this for elements of our standard torus T .
Let m be the biggest integer $\leq \frac{n}{2}$ so that T has m blocks

$$B_1 = \begin{pmatrix} \cos\theta_1 & \sin\theta_1 \\ -\sin\theta_1 & \cos\theta_1 \end{pmatrix} , \ldots, B_m = \begin{pmatrix} \cos\theta_m & \sin\theta_m \\ -\sin\theta_m & \sin\theta_m \end{pmatrix}$$ arranged along

the diagonal. Let $\psi_1 \in T$ be the identity except in block B_i .
Then any element of T is surely a finite product of elements $\psi_1, \ldots, \psi_m$.

Let ϕ_i be reflection in the hyperplane formed by all coordinates except those for B_i and the first coordinate for B_i .

Let ρ_i be reflection in

the hyperplane formed by all coordinates except those for B_i and the coordinate at angle $\frac{\theta_i}{2}$ in the B_i plane. Then $\rho_i \circ \phi_i = \psi_i$.
Thus every element of T is the product of finitely many reflections.

Given $A \in SO(n)$, choose B such that

$$BAB^{-1} \in T \quad .$$

Writing BAB^{-1} as a finite product of reflections and using Proposition 12 gives A as such a product. Finally, let $C \in \Theta(n)$. Then $C = AD$ where $A \in SO(n)$ and D is a reflection (Exercise).

q.e.d.

F. Exercises

1. Show that $A = \begin{pmatrix} \cos \frac{\pi}{4} & \sin \frac{\pi}{4} \\ -\sin \frac{\pi}{4} & \cos \frac{\pi}{4} \end{pmatrix} \in SO(2)$ has no A - stable subspace in R^2 other than $\{o\}$. Find all A - stable subspaces in R^2 for $A = \begin{pmatrix} 0 & 1 \\ 1 & 0 \end{pmatrix}$.

2. Find all eigenvectors and eigenvalues of

$$\begin{pmatrix} 0 & 1 \\ 1 & 0 \end{pmatrix} , \begin{pmatrix} 1 & 0 \\ 0 & 0 \end{pmatrix} , \begin{pmatrix} 1 & 0 \\ 0 & 1 \end{pmatrix} , \begin{pmatrix} 0 & 1 \\ 0 & 1 \end{pmatrix} \quad .$$

3. In $\mathbb{H}^n$ show that if $\langle x, y \rangle$ has zero real part for every y, then $x = 0$.

4. Let $D = \begin{pmatrix} -1 & & & \bigcirc \\ & 1 & & \\ & & \ddots & \\ \bigcirc & & & 1 \end{pmatrix}$. Show that $\Theta(n) - SO(n) = SO(n)D$. Show that D may be replaced by any other element of $\Theta(n) - SO(n)$.

5. Prove that if $\lambda \in \mathbb{H}$ has length 1, there exists $q \in \mathbb{H}$ of length 1 such that $q\lambda\bar{q}$ is a complex number.

6. Let L_1, L_2 be two lines through o in R^2 with angle θ between them. Let ϕ_1 be reflection in L_1 and ϕ_2 be reflection in L_2. Let W be the subgroup of $\mathcal{O}(2)$ generated by ϕ_1 and ϕ_2. For what values of θ will W be a finite group?

7. Recall that in order to prove that

$$1 + \dim SU(n) = \dim U(n)$$

in Chapter IV we used the fact that

$$e^{TrA} = \det e^A \ .$$

Prove this now for $A \in U(n)$. (We can write

$$A = BCB^{-1}$$

where B is unitary and C is an element of our standard maximal torus in $U(n)$. Then

$$e^A = e^{BCB^{-1}} = Be^C B^{-1}$$

and $C = \begin{pmatrix} \lambda_1 & & \bigcirc \\ & \ddots & \\ \bigcirc & & \lambda_n \end{pmatrix}$ with $|\lambda_j| = 1$ so $e^C = \begin{pmatrix} e^{\lambda_1} & & \bigcirc \\ & \ddots & \\ \bigcirc & & e^{\lambda_n} \end{pmatrix} \ .$

Thus $\det e^A = \det (Be^C B^{-1}) = \det e^C = e^{\lambda_1} \ldots e^{\lambda_n} = e^{\lambda_1 + \ldots + \lambda_n}$

$= e^{TrC} = e^{TrA} \ .)$

Chapter 9
Conjugacy of Maximal Tori

A. Monogenic groups

Definition: A subset Y of a space X is said to be <u>dense in</u> X if every nonempty open set in X contains at least one point in Y .

Examples: Both $\mathbb{Q}$ and $\mathbb{R} - \mathbb{Q}$ are dense in $\mathbb{R}$. $\{(x_1, x_2) \in \mathbb{R}^2 \mid x_1, x_2 \in \mathbb{Q}\}$ is dense in $\mathbb{R}^2$.

Lemma: If Y is dense in X and C is a closed set in X with $Y \subset C$, then $C = X$.

Proof: $X - C$ is open, so if it is nonempty it must contain some $y \in Y$. But $y \in X - C$ contradicts $Y \subset C$.

Definition: A matrix group G is <u>monogenic</u> (i.e., one generator) if there exists $x \in G$ such that

$$\{x, x^2, x^3, \ldots\}$$

is dense in G . Then any such x is called a __generator__.

Note that the additive group R is not monogenic, nor are any of the vector groups R^n . Of course, the identity (matrix) cannot ever be a generator. Consider the circle group S^1 thought of as the additive group of reals modulo 1 . If $x = \frac{p}{q} \in S^1$ is rational, then the powers

$$x, 2x, 3x, 4x, \ldots$$

will not form a dense set in S^1 because they will all lie in the set

$$\{\frac{1}{q}, \frac{2}{q}, \ldots, \frac{q-1}{q}, \frac{q}{q} = 0\} \quad .$$

But if we choose x to be irrational we should get a generator. However, we will use a more topological way of finding a generator, and it will generalize immediately to higher dimensional tori.

__Proposition__ 1: S^1 __is__ __monogenic__ (S^1 = __reals__ __mod__ 1) .

__Proof__: Let $U_1, U_2, U_3, \ldots$ be a countable basis for the open sets in S^1 . Then to prove the proposition it suffices to find $x \in S^1$ such that: Given U_k , there exists an $n \in N = \{1, 2, 3, \ldots\}$ such that

$$nx \in U_k .$$

We proceed as follows.

Choose a nondegenerate closed interval $I_1 = [a_1, b_1] \subset U_1$.
Choose an $n_1 \in N$ such that length $[n_1 a_1, n_1 b_1] \geq 2$. Then $n_1 I_1$ is all of S^1 , and thus we can

choose I_2 (nondeg.) such that $n_1 I_2 \subset U_2$,

choose n_2 such that $n_2 I_2$ is all of S^1.

Then we can

choose I_3 (non deg.) such that $n_3 I_3 \subset U_3$,

choose n_3 such that $n_3 I_3$ is all of S^1 , etc.

Because $U_1, U_2, U_3, \ldots$ must contain arbitrarily small sets, we have

that $I_1 \cap I_2 \cap I_3 \cap \ldots$ is a single element x .

But x is a generator for S^1 . For, given U_k , we have

$n_{k-1} x \in n_{k-1} I_k \subset U_k$.   q.e.d.

Now the r-dimensional torus T is just all r-tuples of real

numbers, each taken modulo 1 . So by using cubes instead of intervals

we get

Proposition 2: The r-torus T is monogenic.

B. Conjugacy of maximal tori

We are now ready to prove that for a connected matrix group,

any two maximal tori are conjugate. Let $G \in \{SO(n), U(n), SU(n), Sp(n)\}$

and let T be our standard maximal torus in G . We know that

$$G = \bigcup_{x \in G} x T x^{-1} .$$

Proposition 3: Let T' be any maximal torus in G . Then for

some $x \in G$, $T' = x T x^{-1}$.

Proof: Since T' is a torus it is monogenic and we choose a

generator y for T' . Then for some x

$$y \in xTx^{-1} \ .$$

Now xTx^{-1} is a group, so

$$Y = \{y, y^2, y^3, \ldots\} \subset xTx^{-1} \ ,$$

and, of course, $Y \subset T'$. Thus Y is contained in the closed set $T' \cap xTx^{-1}$ in T' and Y is dense in T' . So by our lemma we have

$$T' \cap xTx^{-1} = T' \quad \text{or} \quad xTx^{-1} \supset T' \ .$$

But xTx^{-1} is a maximal torus in G so $xTx^{-1} = T'$.

Definition: The rank of a matrix group G is the dimension of a maximal torus in G .

This is clearly an invariant for isomorphisms.

C. The isomorphism question again

For a matrix group G we now have two numerical invariants, its dimension n and its rank r , and one subgroup invariant, its center C . Furthermore, we have calculated these and we now tabulate our results according to rank.

Rank 1:	Group	Dimension	Center
	U(1)	1	S^1
	SU(2)	3	$\mathbb{Z}/2$
	SO(2)	1	S^1
	SO(3)	3	$\{I\}$
	Sp(1)	3	$\{I,-I\}=\mathbb{Z}/2$.

Rank 2:			
	U(2)	4	S^1
	SU(3)	8	$\mathbb{Z}/3$
	SO(4)	6	$\{I,-I\}=\mathbb{Z}/2$
	SO(5)	10	$\{I\}$
	Sp(2)	10	$\{I,-I\}$.

Rank 3:			
	U(3)	9	S^1
	SU(4)	15	$\mathbb{Z}/4$
	SO(6)	15	$\{I,-I\}=\mathbb{Z}/2$
	SO(7)	21	$\{I\}$
	Sp(3)	21	$\{I,-I\}=\mathbb{Z}/2$.

For rank 4 and greater we have

$$\dim U < \dim SU < \dim SO(\text{even}) < \dim SO(\text{odd}) = \dim Sp .$$

Since SO(odd) and Sp have different centers, they are not isomorphic. So we just need to look at ranks 1, 2, 3.

We know that $U(1) \simeq SO(2)$ (circle group). $SO(3) \neq Sp(1)$ and we proved in Chapter II that $SP(1) \simeq SU(2)$. So we know all about rank 1.

For rank 2 only SO(5) and Sp(2) have the same dimension
and they have nonisomorphic centers.

For rank 3 SU(4) and SO(6) have the same dimension and
different centers, and the same applies to SO(5) and Sp(2) .
So we have solved our isomorhism problem for these groups. But
we can generate some other groups and the isomorphism question
for these needs to be resolved. In the next section we consider ways
of getting new groups.

If H is any proper subgroup of the center (S^1) of U(n) ,
then we have a group U(n)/H , and H is finite, so

$$\dim \frac{U(n)}{H} \equiv n^2 = \dim U(n)$$

so these groups are not isomorphic to any of our others.

If H is a finite proper subgroup of the center Z/n of
SU(n) (n not a prime), then

$$\dim \frac{SU(n)}{H} = n^2 - 1 = \dim SU(n)$$

and again these are not isomorphic with any of our other groups except
we might have

$$\frac{SU(4)}{Z/2} \cong SO(6) .$$

Indeed, we will see later that this is the case.

D. Simple groups, simply-connected groups

Definition: A matrix group G is simple if it has no nontrivial normal subgroup. This is the same as saying it has no quotient groups other than itself and the trivial group.

If G is not simple, then by choosing a nontrivial normal subgroup H of G we get a new group G/H . So we want to see which of our groups are simple, and, when they are not, what normal subgroups they have. We state the result, but a proof would take us too far afield.

Theorem: $\dfrac{U(n)}{center} \cong \dfrac{SU(n)}{center}$ is simple.

$SO(2n+1)$ is simple.

$\dfrac{SO(2n)}{center}$ is simple.

$\dfrac{Sp(n)}{center}$ is simple.

So we only get the groups we already know about.

Another way of generating new groups is somewhat more sophisticated. A path ω in G is a smooth curve $\omega : [0,1] \to G$ with $\omega(0) = e$. Let $P(G)$ be the set of all paths in G . Then $P(G)$ becomes a group if we define

$$(\omega\sigma)(t) = \omega(t)\sigma(t) .$$

Proposition 4: If G is a connected group, the map $\rho : P(G) \to G$, defined by

$$\rho(\omega) = \omega(1)$$

is a surjective homomorphism.

Proof: G connected implies that for any $x \in G$ there is a path from e to x in G , so ρ is surjective. Now

$$\rho(\omega\sigma) = (\omega\sigma)(1) = \omega(1)\sigma(1) = \rho(\omega)\rho(\sigma) \; ,$$

so ρ is a homomorphism.

We denote the kernel of ρ by $\Omega(G)$ and call it the <u>loops</u> in G .

Definition: G is <u>simply-connected</u> means that $\Omega(G)$ is connected.

If G is not simply-connected, let $\Omega^{\circ}(G)$ be the identity component of $\Omega(G)$. Then $\Omega^{\circ}(G)$ is also a normal subgroup of $P(G)$ and we set

$$\tilde{G} = \frac{P(G)}{\Omega^{\circ}(G)}$$

and call this group the <u>universal covering group</u> of G .

So we get new groups this way whenever G is not simply-connected. Clearly $\rho : P(G) \to G$ induces a homomorphism

$$\rho : \tilde{G} \to G \quad .$$

The kernel of ρ is denoted by $\pi_1(G)$ and is called the <u>fundamental group</u> of G . Clearly G is simply-connected $\Leftrightarrow \pi_1(G) = 1$ (i.e., the trivial group). We state but do not prove the following.

Theorem: $\pi_1(Sp(n)) = 1$ for $n = 1,2,\ldots$

$\pi_1(U(n)) = 1$ for $n = 2,3,\ldots$

$\pi_1(SU(n)) = 1$ for $n = 2,3,\ldots$

$\pi_1(SO(n)) = \frac{Z}{2}$ for $n = 3,4,\ldots$.

So the only new groups we can generate as universal covering groups are

$$\widetilde{SO(n)} .$$

We see that $\rho : \widetilde{SO(n)} \to SO(n)$ is a 2-1 homomorphism. We construct and study these new groups in the next chapter.

E. Exercises

1. Let S^1 be the additive group of reals mod 1 . Show that if $x \in [0,1]$ is irrational, then x is a generator for S^1 . (Hint: For $y = \frac{p}{q}$ rational, $\{ny \mid n = 1,2,\ldots\}$ gets within $\frac{1}{q}$ of each point in S^1 . Show that if x is irrational we can write $x = \lim\limits_{i \to \infty} y_i$ where each $y_i = \frac{p_i}{q_i}$ is rational and $\lim\limits_{i \to \infty} q_i = \infty$.)

2. Let G be a matrix group which is not necessarily connected. Let $G^\circ = \left\{ x \in G \mid \exists \text{ path } \omega : [0,1] \to G \atop \omega(0) = 0 \;\; \omega(1) = x \right\}$.

Call this the identity component of G . Show that G° is a subgroup of G .

3. Let G be a matrix group and H a normal subgroup of G. Show that the identity component H° of H is also a normal subgroup of G .

Chapter 10
Spin(k)

A. Clifford Algebras

One way of constructing groups which are subsets of some R^n is: Let G be a finite-dimensional real algebra and let G be the group of units in G. We get more groups as subgroups of G. For example, we have used the algebra $M_n(R)$ in which the group of units is $GL(n,R)$ and we have the important subgroup $SO(n)$. Our groups $Spin(k)$ are subgroups of the group of units in the Clifford algebra C_k.

For $k = 0,1,2,\ldots$ we will define a real algebra C_k of dimension 2^k. First we set

$$C_0 = R .$$

Next, C_1 is to be two dimensional: we take a basis $1,e$; let 1 act as a multiplicative identity and define $e^2 = -1$. This specifies the multiplication

$$(a + be)(c + de) = (ac - bd) + (ad + bc)e$$

and we see that

$$C_1 \cong C$$

as 2-dimensional real algebras.

Next C_2 is to be 4-dimensional. We take a basis

$$1 \ , \ e_1 \ , \ e_2 \ , \ e_1 e_2$$

and set $e_1^{\ 2} = -1$, $e_2^{\ 2} = -1$ and $e_2 e_1 = -e_1 e_2$. This specifies

the multiplication and we see that the assignment

$$1 \mapsto 1$$
$$e_1 \mapsto i$$
$$e_2 \mapsto j$$
$$e_1 e_2 \mapsto k$$

gives an isomorphism

$$C_2 \cong \mathbb{H} \ .$$

__Definition__: C_k is the algebra which is generated, as an

algebra, by $e_1, e_2, \ldots, e_k$ subject to

$$e_i^{\ 2} = -1 \quad \text{and} \quad e_j e_i = -e_i e_j \quad \text{if } i \neq j \ .$$

To explain what is meant by "generated as an algebra," consider

C_3 . To e_1, e_2, e_3 we must add 1 and products of the e_i's to get

a vector-space basis for C_3 , and we get

$$\{1, e_1, e_2, e_3, e_1 e_2, e_1 e_3, e_2 e_3, e_1 e_2 e_3\}$$

which gives dimension $2^3 = 8$. The same argument shows that

$$\{e_{i_1} \cdots e_{i_r} \mid i_1 < i_2 < \ldots < i_r \ , \ 0 \leq r \leq k\}$$

is a vector-space basis for C_k ($r = 0$ denotes the element 1 of the basis).

Note that sending e_i in C_{k-1} to e_i in C_k imbeds C_{k-1} as a subalgebra of C_k ,

$$C_0 \subset C_1 \subset \cdots \subset C_{k-1} \subset C_k \subset \cdots \quad .$$

Now C_0, C_1 are fields and C_2 is a skew field but we cannot expect this for $k > 2$.

Proposition 1: For all $k > 2$, C_k has divisors of zero.

Proof: It suffices to find a divisor of zero in C_3 ; i.e., an $x \neq 0$ such that there exists a $y \neq 0$ with $xy = 0$. Well,

$$(1 + e_1 e_2 e_3)(1 - e_1 e_2 e_3) = 1 - e_1 e_2 e_3 e_1 e_2 e_3 = 0 \; .$$

There are two notable ways of getting algebras which have divisors of zero.

Definition: If $\mathbb{G}$ and $\mathbb{B}$ are real algebras, their direct sum $\mathbb{G} \oplus \mathbb{B}$ is the set of all pairs (x,y) with $x \in \mathbb{G}$ and $y \in \mathbb{B}$ with operations

$$(x,y) + (z,w) = (x+z, y+w)$$

$$r(x,y) = (rx, ry)$$

$$(x,y)(z,w) = (xz, yw) \quad .$$

Proposition 2: $\mathbb{G} \oplus \mathbb{B}$ has divisors of zero, and for any field F , $M_n(F)$ has divisors of zero if $n \geq 2$.

<u>Proof</u>: $(1,0)(0,1) = (0,0)$

$$\begin{pmatrix} 1 & 1 \\ 1 & 1 \end{pmatrix} \begin{pmatrix} 1 & -1 \\ -1 & 1 \end{pmatrix} = \begin{pmatrix} 0 & 0 \\ 0 & 0 \end{pmatrix} \ .$$

For our purposes we need to know the Clifford algebras up through C_5 . There is a uniform way of getting up through C_4 (due to Alan Wiederhold). We consider

$$\mathbb{C} \subset \mathbb{H} \subset \mathbb{H} \oplus \mathbb{H} \subset M_2(\mathbb{H})$$

as follows: $\alpha \in \mathbb{C}$ is represented by $\begin{pmatrix} \alpha & 0 \\ 0 & \alpha \end{pmatrix}$, $q \in \mathbb{H}$ by $\begin{pmatrix} q & 0 \\ 0 & q \end{pmatrix}$ and $(q_1, q_2) \in \mathbb{H} \oplus \mathbb{H}$ by $\begin{pmatrix} q_1 & 0 \\ 0 & q_2 \end{pmatrix}$. Then the assignment

$$\begin{pmatrix} i & 0 \\ 0 & i \end{pmatrix} \to e_1$$

$$\begin{pmatrix} j & 0 \\ 0 & j \end{pmatrix} \to e_2$$

$$\begin{pmatrix} k & 0 \\ 0 & -k \end{pmatrix} \to e_3$$

$$\begin{pmatrix} 0 & k \\ k & 0 \end{pmatrix} \to e_4$$

gives isomorphisms

$$\mathbb{C} \xrightarrow{\approx} C_1$$

$$\mathbb{H} \xrightarrow{\approx} C_2$$

$$\mathbb{H} \oplus \mathbb{H} \xrightarrow{\approx} C_3$$

$$M_2(\mathbb{H}) \xrightarrow{\approx} C_4 \ .$$

Finally, the assignment

$$\begin{pmatrix} i & 0 & 0 & 0 \\ 0 & -i & 0 & 0 \\ 0 & 0 & i & 0 \\ 0 & 0 & 0 & -i \end{pmatrix} \quad \begin{pmatrix} 0 & -1 & 0 & 0 \\ 1 & 0 & 0 & 0 \\ 0 & 0 & 0 & -1 \\ 0 & 0 & 1 & 0 \end{pmatrix} \quad \begin{pmatrix} 0 & i & 0 & 0 \\ i & 0 & 0 & 0 \\ 0 & 0 & 0 & -i \\ 0 & 0 & -i & 0 \end{pmatrix}$$

$$\downarrow \qquad\qquad\qquad \downarrow \qquad\qquad\qquad \downarrow$$

$$e_1 \qquad\qquad\qquad e_2 \qquad\qquad\qquad e_3$$

$$\begin{pmatrix} 0 & 0 & 0 & -1 \\ 0 & 0 & -1 & 0 \\ 0 & 1 & 0 & 0 \\ 1 & 0 & 0, & 0 \end{pmatrix} \quad \begin{pmatrix} 0 & 0 & 0 & -i \\ 0 & 0 & -i & 0 \\ 0 & -i & 0 & 0 \\ -i & 0 & 0 & 0 \end{pmatrix}$$

$$\downarrow \qquad\qquad\qquad \downarrow$$

$$e_4 \qquad\qquad\qquad e_5$$

induces an isomorphism $M_4(c) \overset{\Downarrow}{\to} C_5$.

B. Pin(k) and Spin(k)

Let R^k denote the k-space in C_k spanned by $e_1, \ldots, e_k$ and let S^{k-1} be the unit sphere in R^k .

Proposition 3: If C^{*}_k denotes the group of units in C_k , then $S^{k-1} \subset C^{*}_k$; i.e., each $x \in S^{k-1}$ is a unit.

Proof: Let $x = a_1 e_1 + \ldots + a_k e_k$ with $a_1^2 + \ldots + a_k^2 = 1$. Then

$$(a_1 e_1 + \ldots + a_k e_k)((-a_1)e_1 + \ldots + (-a_k)e_k) = a_1^2 + \ldots + a_k^2 = 1 .$$

Definition: Pin(k) is the subgroup of C^{*}_k generated by S^{k-1} .

Thus each element of $\text{Pin}(k)$ is a finite product of elements of S^{k-1}.

We define a __conjugation__ in C_k. It suffices to define it on basis elements and we set

$$(e_{i1} \cdots e_{ir})^* = (-e_{ir}) \cdots (-e_{i1}) = (-1)^r e_{ir} \cdots e_{i1} \quad .$$

For example,

$$1^* = 1 \ , \quad e_i^* = -e_i \ , \quad (e_i e_j)^* = -e_i e_j \ ,$$

$$(e_i e_j e_k)^* = e_i e_j e_k \ , \quad \text{etc.}$$

Clearly $(x^*)^* = x$. Also, we have $(xy)^* = y^* x^*$ because

$$((e_{i1} \cdots e_{ir})(e_{j1} \cdots e_{js}))^* = (-e_{js}) \cdots (-e_{j1})(-e_{ir}) \cdots (-e_{i1})$$

$$= (e_{j1} \cdots e_{js})^* (e_{i1} \cdots e_{ir})^* \quad .$$

We also define an __automorphism__ α of C_k by $\alpha(e_i) = -e_i$. (Note that conjugation is not an automorphism, because it agrees with α on each e_i, but $\alpha(e_i e_j) = (-e_i)(-e_j) = e_i e_j$ whereas $(e_i e_j)^* = -e_i e_j$.)

__Definition:__ For $u \in \text{Pin}(k)$ and $x \in R^k$ we set

$$\rho(u)(x) = \alpha x u^* \quad .$$

It is not clear that $\rho(u)(x) \in R^k$ but it is a consequence of the next two propositions.

__Proposition 4:__ __If__ $u \in S^{k-1} \subset \text{Pin}(k)$, __then__ $\rho(u)$ __is__ __reflection in__ R^k __in the hyperplane perpendicular to__ u .

Proof: Pick an orthonormal basis $u_1, \ldots, u_k$ of R^k with $u_1 = u$. Consider

$$\rho(u_1)u_i = \alpha(u_1)u_i u_1^* = (-u)u_i(-u_1) = u_1 u_i u_1 .$$

If $i \neq 1$ this equals u_i and if $i = 1$ it equals $-u_1$, proving the proposition.

Proposition 5: ρ is a homomorphism of Pin(k) onto $\mathcal{O}(k)$ and ker $\rho = \{1,-1\}$.

Proof: Every element of Pin(k) is a finite product of elements of S^{k-1} , so to prove ρ is a homomorphism it suffices to take $u,v \in S^{k-1}$ and $x \in R^k$. Then

$$\rho(uv)(x) = \alpha(uv)x(uv)^* = \alpha(u)(\alpha(v)xv^*)u^* = \rho(u)(\rho(v)x) .$$

It now follows that for each $u \in$ Pin(k) , $\rho(u)$ does map R^k into R^k and it is orthogonal since it is a product of reflections. In Chapter VII we showed that $\mathcal{O}(k)$ is generated by reflections so that ρ is surjective.

It is clear that 1 and -1 are in the kernel of ρ . If we have $e_{i1} \cdots e_{ir}$ in the kernel of ρ with $r > 1$ (we cannot have $r = 1$) we get a contradiction as follows: If $\rho(e_{i1} \cdots e_{ir})$ is the identity, then for every $x \in R^k$ we must have

$$\alpha(e_{i1} \cdots e_{ir})x(e_{i1} \cdots e_{ir})^* = x \quad \text{or}$$

$$e_{i1} \cdots e_{ir} x = x e_{i1} \cdots e_{ir} .$$

Taking $x = e_{i1}$ gives $(-1)^{r-1} e_{i1} e_{i1} \cdots e_{ir} = e_{i1} e_{i1} \cdots e_{ir}$ and this gives $e_{i2} \cdots e_{ir} = 0$ since r must be even (the product of an odd number of reflections cannot be the identity).

Definition: $\text{Spin}(k) = \rho^{-1}(\text{SO}(k))$.

Let us calculate $\text{Spin}(1)$, $\text{Spin}(2)$ and $\text{Spin}(3)$. $C_1 = \mathbb{C}$ so $C_1^* = \mathbb{C} - (0)$ and $\text{Pin}(1)$ is the subgroup of C_1^* generated by $S^0 = \{e_1, -e_1\}$. This is just $\{e_1, e_1^2 = -1, e_1^3 = -e_1, e_1^4 = 1\} \cong \frac{\mathbb{Z}}{4}$ Now $\rho(e_1) = \rho(-e_1)$ is a reflection and

$$\text{Spin}(1) = \{1, -1\} \cong \mathbb{Z}/2 \ .$$

$C_2 = \mathbb{H}$ so $C_2^* = \mathbb{H} - (0)$ and $\text{Pin}(2)$ is the subgroup generated by

$$S^1 = \{ae_1 + be_2 \mid a^2 + b^2 = 1\} \ .$$

Now $(ae_1 + be_2)(ce_1 + de_2) = -(ac + bd) + (ad - bc)e_1 e_2$, so that $\text{Pin}(2)$ must contain

$$S = \{c + de_1 e_2 \mid c^2 + d^2 = 1\} \ .$$

Actually it is easy to see that

$$\text{Pin}(2) = S \cup S^1 \quad \text{and} \quad \text{Spin}(2) = S \ .$$

Similar computations easily give

$$\text{Spin}(3) = \{a + be_1 e_2 + ce_1 e_3 + de_2 e_3 \mid a^2 + b^2 + c^2 + d^2 = 1\}$$

and we see that

$$\text{Spin}(3) \cong \text{Sp}(1) \ .$$

We conclude this section by calculating the centers of the

spin groups.

First off we always have $\{1,-1\}$ in the center. This is all of the center in Spin(1) (being all of Spin(1)) and Spin(2) equals its center. So we now assume

$$k \geq 3 .$$

Suppose $e_{i1}\cdots e_{ir}$ fails to contain e_j . Consider

$$(e_{i1}\cdots e_{ir})(e_{ir}e_j) = (-1)^{r-1}e_{ir}(e_{i1}\cdots e_{ir})e_j$$

$$= (-1)^{r-1}(-1)^r(e_{ir}e_{ij})(e_{i1}\cdots e_{ir})$$

$$= -(e_{ir}e_{ij})(e_{i1}\cdots e_{ir}) ,$$

proving $e_{i1}\cdots e_{ir}$ is not in the center. Thus $e_1 e_2 \cdots e_k$ and $-e_1 e_2 \cdots e_k$ are the only candidates (besides 1 and -1). Now for k odd $e_1 \cdots e_k$ is not in Spin(k) . For $k = 2n$ we see that

$$(e_1 e_2 \cdots e_{2n})(e_{i1}\cdots e_{ir}) \quad (\text{ r even })$$

$$= (e_{i1}\cdots e_{ir})(e_1 e_2 \cdots e_{2n}) = (-1)^{(2n-1)r} = 1$$

so that $e_1 \cdots e_{2n}$ is in the center.

Proposition 6: (i) If k is odd Center Spin(k) = $\{1,-1\}$;
(ii) Center Spin(2n) = $\mathbb{Z}/4$ if n is odd;
(iii) Center Spin(2n) = $\mathbb{Z}/2 \oplus \mathbb{Z}/2$ if n is even.

Proof: We have proved (i) and we have seen that
Center Spin(2n) = $\{1,-1,e_1\cdots e_{2n},-e_1\cdots e_{2n}\}$. Note that
$(e_1\cdots e_{2n})(e_1\cdots e_{2n}) = (-1)^{n(2n+1)}$ which is 1 if n is even and

is -1 if n is odd. This proves the proposition.

C. The isomorphisms

We have already noted that if $e_{i1} \cdots e_{ir} \in \mathrm{Spin}(k)$ then r is even. In this case we have

$$(e_{i1} \cdots e_{ir})(e_{i1} \cdots e_{ir})^* = 1 .$$

So that if $x \in \mathrm{Spin}(k)$, x is a linear combination of "even-graded" basis elements and $xx^* = 1$.

Definition: Define $\phi : C_{k-1} \to C_k$ by

$$\phi(e_i) = e_i e_k \quad \text{for} \quad i = 1, \ldots, k-1 .$$

Then ϕ extends to an algebra homomorphism since $(e_i e_k)^2 = -1$ and $(e_i e_k)(e_j e_k) = -(e_j e_k)(e_i e_k)$. Indeed, $e_i e_k e_j e_k = e_i e_j$ and we see that ϕ is an isomorphism of C_{k-1} onto the even-graded part of C_k .

Proposition 7: The isomorphisms ψ (§A) and ϕ satisfy:

$$(\phi(x))^* = \phi(x^*)$$

$$\psi(^t \bar{M}) = (\psi(M))^* .$$

Proof: $(\phi(e_i))^* = (e_i e_k)^* = (-e_k)(-e_i) = -e_i e_k = \phi(e_i^*)$.

For C_1, C_2, C_3, C_4 we check the second formula on the basic matrices

$$M = \begin{pmatrix} i & 0 \\ 0 & i \end{pmatrix} \overset{\psi}{\downarrow} e_1 \qquad\qquad {}^t\bar{M} = \begin{pmatrix} -i & 0 \\ 0 & -i \end{pmatrix} \overset{\psi}{\downarrow} -e_1 \qquad .$$

$$\parallel$$

$$(\psi(M))^*$$

The others work the same way, and the verification for $M_4(\mathbb{C}) \overset{\psi}{\to} C_5$ is equally routine.

Theorem: The composed maps

$$Sp(2) \to M_2(\mathbb{H}) \overset{\psi}{\underset{\cong}{\to}} C_4 \overset{\phi}{\cong} C_5$$

$$SU(4) \to M_4(\mathbb{C}) \overset{\psi}{\underset{\cong}{\to}} C_5 \overset{\phi}{\cong} C_6$$

give isomorphism of $Sp(2)$ with $Spin(5)$ and $SU(4)$ with $Spin(6)$.

Proof: $Sp(2)$ and $SU(4)$ are those matrices such that ${}^t\bar{M} = I$ so that

$$1 = \psi(M {}^t\bar{M}) = \psi(M)\psi(M)^* \quad \text{and}$$

$$1 = \phi\psi(M)\phi(\psi(M)^*) = \phi\psi(M)(\phi\psi(M))^* \quad .$$

We get all even graded units x with $xx^* = 1$ in this manner. Thus

$$Spin(5) \subset \phi\psi(Sp(2))$$

$$Spin(6) \subset \phi\psi(SU(4)) ,$$

and $\phi\psi(Sp(2))$ is a closed connected manifold of dimension 10 and $Spin(5)$ is a closed manifold of dimension 10 , so $Spin(5) = \phi\psi(Sp(2))$. Similarly, the two 15-dimensional manifolds $Spin(6)$ and $\phi\psi(SU(4))$ are equal.

144

D. Exercises

1. Show that

$$U(1) \to \mathbb{C} \overset{\Downarrow}{\cong} C_1 \overset{\Phi}{\cong} C_2$$

gives an isomorphism $U(1) \cong \mathrm{Spin}(2)$.

2. Show that

$$Sp(1) \to \mathbb{H} \overset{\Downarrow}{\cong} C_2 \overset{\Phi}{\cong} C_3$$

gives an isomorphism $Sp(1) \cong \mathrm{Spin}(3)$.

3. Show that

$$Sp(1) \times Sp(1) \to \mathbb{H} \oplus \mathbb{H} \overset{\Downarrow}{\cong} C_3 \overset{\Phi}{\cong} C_4$$

gives an isomorphism $Sp(1) \times Sp(1) \cong \mathrm{Spin}(4)$.

4. Let G be an abelian group with exactly four elements. Show that $G \cong \mathbb{Z}/4$ if and only if G has an element of order 4 and $G \cong \mathbb{Z}/2 \oplus \mathbb{Z}/2$ otherwise.

Chapter 11
Normalizers, Weyl Groups

A. Normalizers

We saw in Chapter X that $Sp(1) \cong Spin(3)$ and $Sp(2) \cong Spin(5)$ (also that $Spin(6) \cong SU(4)$). The only remaining question in our isomorphism puzzle is whether $Sp(k)$ and $Spin(2k+1)$ are isomorphic for $k \geq 3$. In this chapter, we will use normalizers and Weyl groups to show that they are not.

Definition: Let S be a nonempty subset of a group G . Set

$$N(S) = \{x \in G \mid xSx^{-1} = S\}$$

and call $N(S)$ the normalizer (of S in G). $N(S)$ is a subgroup of G (Exercise 5, Chapter I).

Lemma 1: If S is a subgroup of G , then S is a normal subgroup of $N(S)$.

Proof: (Exercise 5, Chapter I).

We calculate now the normalizers of our standard maximal tori in S^3 ($= Sp(1)$) and $SO(3)$.

Let x be an element of a group G , and define

$$A_x : G \to G$$

by $A_x(g) = xgx^{-1}$ (conjugation by x). Then A_x is an isomorphism of G onto G.

Our standard maximal torus in $Sp(1)$ is $T = \{a+ib \mid a^2+b^2 = 1\}$. Let N be the normalizer of T in $Sp(1)$. If $q \in N$ then $A_q : T \to T$ is an isomorphism of the circle group T onto itself.

Lemma 2: There are exactly two isomorphisms of the circle group.

Proof: Let S^1 (as T above) be the set of complex numbers of unit length. Then $F = \{1,-1,i,-i\}$ is the set of all 4th roots of unity in S^1. If $\phi : S^1 \to S^1$ is an isomorphism, it must have $\phi(1) = 1$, and therefore $\phi(-1) = -1$. Thus $\phi(i) \in \{i,-i\}$. Both possibilities give isomorphisms and knowing ϕ on F determines ϕ. (See Exercise #5.)   q.e.d.

Let $q \in N$. If A_q is the identity then q commutes with each $x \in T$, so $q \in T$. So suppose A_q is the other isomorphism. Since $|q| = 1$, $q^{-1} = \bar{q}$, so we have

$$q i \bar{q} = -i .$$

If we write $q = x + iy + jz + kw$, this gives

$$x^2 + y^2 - z^2 - w^2 = -1 ,$$

and since

$$x^2 + y^2 + z^2 + w^2 = 1$$

we get $x = 0 = y$. Furthermore, if $q = jz + kw$ with $z^2 + w^2 = 1$

we do get $A_q(i) = -i$. Hence

$$N = T \cup T^* \quad \text{(disjoint union)}$$

where $T^* = \{jz + kw \mid z^2 + w^2 = 1\}$. Thus the quotient group
$N/T = \mathbb{Z}/2$ (T is normal in N), and we have

$$T \to N \to \mathbb{Z}/2 .$$

Next we calculate the normalizer N' of the standard maximal
torus T' in SO(3) . Recall that

$$T' = \left\{ \begin{pmatrix} \cos \theta & \sin \theta & 0 \\ -\sin \theta & \cos \theta & 0 \\ 0 & 0 & 1 \end{pmatrix} \right\} .$$

Now $T' \subset N'$ and we seek the elements of $N' - T'$.

Lemma 3: $N' - T' = \left\{ \begin{pmatrix} \cos \phi & \sin \phi & 0 \\ \sin \phi & -\cos \phi & 0 \\ 0 & 0 & -1 \end{pmatrix} \right\} .$

Proof: Each such matrix is symmetric and is its own inverse
and has determinant +1 , so it is in SO(3) . It is a routine,
but instructive, exercise to show that

$$\begin{pmatrix} \cos \phi & \sin \phi & 0 \\ \sin \phi & -\cos \phi & 0 \\ 0 & 0 & -1 \end{pmatrix} \begin{pmatrix} \cos \theta & \sin \theta & 0 \\ -\sin \theta & \cos \theta & 0 \\ 0 & 0 & 1 \end{pmatrix} \begin{pmatrix} \cos \phi & \sin \phi & 0 \\ \sin \phi & -\cos \phi & 0 \\ 0 & 0 & -1 \end{pmatrix} = \begin{pmatrix} \cos \theta & -\sin \theta & 0 \\ \sin \theta & \cos \theta & 0 \\ 0 & 0 & 1 \end{pmatrix} .$$

It is less routine (but also instructive) to show that these are the
only matrices in SO(3) which give the nontrivial isomorphism
of T' .

So we have:

$$T \to N \to \mathbb{Z}/2 \quad \text{for} \quad Sp(1)$$

$$T' \to N' \to \mathbb{Z}/2 \quad \text{for} \quad SO(3)$$

and both T and T' are circle groups.

Proposition 1: $N' \not\cong N$.

Proof: If $\phi : N \to N'$ were an isomorphism it would have to send T to T' and $N - T$ to $N' - T'$. If $x \in N - T$ we would have $\phi(x^2) = (\phi(x))^2$. But it is easy to check that if $x \in N - T$, then $x^2 = -1$ and if $y \in N' - T'$, then $y^2 = I$.

So we see that N and N' are groups having circle subgroups and isomorphic quotient groups (namely $\mathbb{Z}/2$), but N and N' are not isomorphic. They are <u>different</u> <u>extensions</u> of S^1 by $\mathbb{Z}/2$.

Definition: Let A be an abelian normal subgroup of B and C be the quotient group,

$$A \stackrel{\alpha}{\to} B \stackrel{\beta}{\to} C$$

α being the inclusion and β being the quotient map. We say say this extension <u>splits</u> if there is a homomorphism $\gamma : C \to B$ such that $\beta \circ \gamma$ is the identity on C . (γ is called a homomorphic <u>cross</u> <u>section</u>.)

Then $T' \to N' \to \mathbb{Z}/2$ splits (send identity to identity and send the non identity element of $\mathbb{Z}/2$ to $\begin{pmatrix} 1 & 0 & 0 \\ 0 & -1 & 0 \\ 0 & 0 & -1 \end{pmatrix}$), but $T \to N \to \mathbb{Z}/2$ does not split (because we must send identity to identity and the non identity element of $\mathbb{Z}/2$ into $N - T$ and the result is not a homomorphism).

This is the way we are going to distinguish $Sp(k)$ and $Spin(2k+1)$ for $k \geq 3$.

3. Weyl groups

Let T be a maximal torus in a matrix group G and N be the normalizer of T . We have

$$T \overset{\alpha}{\to} N \overset{\beta}{\to} W$$

where $W = N/T$, α is the inclusion and β is the quotient map.

Definition: W is called the Weyl group of G . We note that if we used another maximal torus T' , then $T' = xTx^{-1}$ and $A_x : G \to G$ would induce an isomorphism of T to T' , N to the normalizer N' of T' and hence induce an isomorphism of Weyl groups (Exercise). Similarly, isomorphic groups will have isomorphic Weyl groups.

Proposition 2: Let $x, y \in N$. Then A_x and A_y are the same isomorphism on $T \Leftrightarrow x$ and y belong to the same coset of T in N .

Proof: $\Leftarrow$

$x = y\tau$ for some $\tau \in T$. So $A_x(t) = xtx^{-1} = y\tau t(y\tau)^{-1} = y\tau t \tau^{-1} y^{-1}$, and since T is abelian this is just $yty^{-1} = A_y(t)$.

$\Rightarrow$

If $xtx^{-1} = yty^{-1}$ for all $t \in T$, then $(y^{-1}x)t(y^{-1}x)^{-1} = t$ so that $y^{-1}x$ commutes with all $t \in T$ and T is maximal so $y^{-1}x \in T$.

Corollary: W <u>acts on</u> T .

<u>Proof</u>: For $w \in W$ choose $x \in \beta^{-1}(w)$ and define $w(t) = A_x(t)$
By Proposition 2, this is independent of the choice of x .

Proposition 3: W <u>is a finite group</u>.

<u>Proof</u>: By the results above it suffices to show that a torus
T has only finitely-many distinct isomorphisms. Now
$T = S^1 \times \dots \times S^1$ and an isomorphism $\phi : T \to T$ is determined by its
restrictions to the coordinate circle groups. But $\phi | S^1$ is deter-
mined by knowing it on the set of <u>fourth roots of unity</u> $(1, i, -1, -i)$
in S^1 . It follows that an isomorphism ϕ of T is determined by
its values on the 4^r (dim T = r) fourth roots of unity. But an
isomorphism sends a fourth root of unity to a fourth root of unity.
So each ϕ is determined by its permutation of 4^r elements and
so the automorphism group of T is finite.

From this it follows that N is an extension of T by a finite
group. In particular then, T is the connected component of N
containing the identity. We have

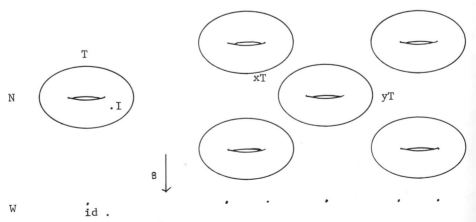

C. Spin(2n+1) <u>and</u> Sp(n)

An isomorphism of Spin(2n+1) onto Sp(n) would induce an isomorphism of their centers and would induce

$$\frac{\text{Spin}(2n+1)}{\text{center}} \cong \frac{\text{Sp}(n)}{\text{center}} \quad .$$

<u>Theorem</u>: <u>For</u> n = 3,4,5,... <u>the normalizer of a maximal torus</u> <u>in</u> $\frac{\text{Spin}(2n+1)}{\text{center}}$ <u>splits, but the normalizer of a maximal torus in</u> $\frac{\text{Sp}(n)}{\text{center}}$ <u>does not split</u>. <u>So for</u> n = 3,4,5,...

$$\text{Spin}(2n+1) \not\cong \text{Sp}(n) \quad .$$

We will prove slightly more; namely,

(*) The normalizer in Sp(n) does not split for any n ;

(†) The normalizer in $\frac{\text{Sp}(n)}{\text{center}}$ splits $\Leftrightarrow$ n $\in$ {1,2} ;

(⏀) The normalizer in $\frac{\text{Spin}(2n+1)}{\text{center}}$ (= SO(2n+1)) splits for n = 1,2,3,... .

<u>Proof</u> <u>of</u> (*): We want to show that

$$T \xrightarrow{\alpha} N \underset{\overset{\leftarrow}{\gamma}}{\overset{\beta}{\longrightarrow}} W$$

does not split in Sp(n) ; i.e. that no homomorphic cross-section γ can exist. We will suppose γ exists and obtain a contradiction.

Let $\eta = \begin{pmatrix} j_1 & & \bigcirc \\ & \ddots & \\ \bigcirc & & 1 \end{pmatrix}$ $\in$ Sp(n) . We claim that $\eta \in N$. Take any

$$t \in \begin{pmatrix} e^{i\theta_1} & & \bigcirc \\ & \ddots & \\ \bigcirc & & e^{i\theta_n} \end{pmatrix} \quad \text{in} \quad T \quad \text{and we calculate}$$

$$\eta t \eta^{-1} = \begin{pmatrix} j_1 & & \bigcirc \\ & \ddots & \\ \bigcirc & & 1 \end{pmatrix} \begin{pmatrix} e^{i\theta_1} & & \bigcirc \\ & \ddots & \\ \bigcirc & & e^{i\theta_n} \end{pmatrix} \begin{pmatrix} -j_1 & & \bigcirc \\ & \ddots & \\ \bigcirc & & 1 \end{pmatrix}$$

$$= \begin{pmatrix} je^{i\theta_1}(-j) & & & \bigcirc \\ & e^{i\theta_2} & & \\ & & \ddots & \\ \bigcirc & & & e^{i\theta_n} \end{pmatrix}$$

and $je^{i\theta_1}(-j) = j(\cos\theta_1 + i \sin\theta_1)(-j) = e^{-i\theta_1}$, so that $\eta t \eta^{-1}$
is again in T .

Let $w = \beta(\eta)$. Then, since $\eta^2 = \begin{pmatrix} -1_1 & \bigcirc \\ \bigcirc & \ddots \\ & & 1 \end{pmatrix} \in T$ we have that

$w^2 = \beta(\eta^2) = 1 \in W$. On the other hand, since γ is assumed to be a
cross section we have

$$\eta' = \gamma(w)$$

must be in the same coset of T as η is, and since γ is to be a
homomorphism we must have

$$(\eta')^2 = \gamma(w^2) = \gamma(1) = I \quad .$$

So we can complete our proof of (*) by showing that no such η'
exists; i.e. no element in the coset of η square to the identity.

Any such η' can be written $\eta' = \eta t$

$$
\eta' = \begin{pmatrix} j_1 & & \\ & \ddots & \\ & & \cdot 1 \end{pmatrix} \begin{pmatrix} e^{i\theta_1} & & \\ & \ddots & \\ & & e^{i\theta_n} \end{pmatrix}
$$

$$
\eta' = \begin{pmatrix} je^{i\theta_1} & & & \bigcirc \\ & e^{i\theta_2} & & \\ & & \ddots & \\ \bigcirc & & & e^{i\theta_n} \end{pmatrix} \quad .
$$

Thus

$$
(\eta')^2 = \begin{pmatrix} je^{i\theta_1}je^{i\theta_1} & & & \bigcirc \\ & e^{2i\theta_2} & & \\ & & \ddots & \\ \bigcirc & & & e^{2i\theta_n} \end{pmatrix}
$$

$$
= \begin{pmatrix} -1 & & & \bigcirc \\ & e^{2i\theta_2} & & \\ & & \ddots & \\ \bigcirc & & & e^{2i\theta_n} \end{pmatrix} \neq I
$$

and (*) is proved.

<u>Proof</u> <u>of</u> (+): We want to prove that the normalizer in $\frac{Sp(n)}{center}$ splits $\Leftrightarrow n \in \{1,2\}$. For $n = 1$, $\frac{Sp(1)}{center} = SO(3)$ and we have seen that N splits for $SO(3)$ (§A) . For $n = 2$ we have $Sp(2) \simeq Spin(5)$ and $\frac{Sp(2)}{center} \simeq SO(5)$ and we will show in (T) that $SO(odd)$ always splits. So it suffices to show here that $\frac{Sp(n)}{center}$

does not split for $n \geq 3$.

For notational convenience we will prove this for $n = 3$ and the general argument will be clear. The idea is like for the proof of (*). We need to find a relation in W which cannot lift (via γ) to a relation in N .

Let $\eta = \begin{pmatrix} j & 0 & 0 \\ 0 & 1 & 0 \\ 0 & 0 & 1 \end{pmatrix}$ and $\sigma = \begin{pmatrix} 0 & 1 & 0 \\ 1 & 0 & 0 \\ 0 & 0 & 1 \end{pmatrix}$ and one checks easily

that η, σ are in N . Let

$$w_1 = \beta(\eta) \quad \text{and} \quad w_2 = \beta(\sigma) \quad .$$

Also since η^2 and σ^2 are in T we have

$$w_1^2 = 1 \quad \text{and} \quad w_2^2 = 1 \quad .$$

Also

$$\eta\sigma = \begin{pmatrix} 0 & j & 0 \\ 1 & 0 & 0 \\ 0 & 0 & 1 \end{pmatrix} \quad .$$

It is left as an exercise to show that the fourth power of $\eta\sigma$ lies in T . So we have

$$(w_1 w_2)^4 = 1 \quad .$$

If we had a homomorphic cross section then we would have to have

$$(\gamma(w_1))^2 \in \{I, -I\} \quad (= \text{center})$$

$$(\gamma(w_2))^2 \in \{I, -I\}$$

$$(\gamma(w_1)\gamma(w_2))^4 \in \{I, -I\} \quad .$$

So we will assume that we have η' in the coset of η, σ' in the oset of σ such that $(\eta')^2$, $(\sigma')^2$ and $(\eta'\sigma')^4$ are all in $\{I,-I\}$.

__Lemma a:__ $(\eta')^2 = -I$.

__Proof:__ We have

$$\eta' = \eta t = \begin{pmatrix} je^{i\theta_1} & 0 & 0 \\ 0 & e^{i\theta_2} & 0 \\ 0 & 0 & e^{i\theta_3} \end{pmatrix}$$

nd $(\eta')^2$ cannot be I since it has -1 in the $1,1$ position.

Thus $(\eta')^2 = -I$ and $\eta' = \begin{pmatrix} je^{i\theta_1} & 0 & 0 \\ 0 & \pm i & 0 \\ 0 & 0 & \pm i \end{pmatrix}$.

__Lemma b:__ For some positive integer m , σ' may be written as

$$\sigma' = \begin{pmatrix} 0 & (-1)^m e^{i\phi} & 0 \\ e^{i\phi} & 0 & 0 \\ 0 & 0 & i^m \end{pmatrix}$$.

__Proof:__ We have

$$\sigma' = \sigma t = \begin{pmatrix} 0 & 1 & 0 \\ 1 & 0 & 0 \\ 0 & 0 & 1 \end{pmatrix} \begin{pmatrix} e^{i\phi_1} & 0 & 0 \\ 0 & e^{i\phi_2} & 0 \\ 0 & 0 & e^{i\phi_3} \end{pmatrix}$$

$$\sigma' = \begin{pmatrix} 0 & e^{i\phi_2} & 0 \\ e^{i\phi_1} & 0 & 0 \\ 0 & 0 & e^{i\phi_3} \end{pmatrix}$$.

Then $(\sigma')^2 = \begin{pmatrix} e^{i(\phi_1+\phi_2)} & 0 & 0 \\ 0 & e^{i(\phi_1+\phi_2)} & 0 \\ 0 & 0 & e^{2i\phi_3} \end{pmatrix}$

and this must be I or -I . This shows that $e^{2i\phi_3}$ is 1 or -1
so the $e^{i\phi_3} = i^m$ for some m . If m is odd $(\sigma')^2$ must be -I
and then $e^{i\phi_1}e^{i\phi_2} = -1$ shows we can take $\phi = -\phi_1$. If m is even
we can take $\phi = \phi_1$ and $e^{i\phi_2} = (-1)^m e^{i\phi}$, and Lemma b is proved.
We finish (†) by

Lemma c: $(\eta'\sigma')^4 = \begin{pmatrix} -1 & 0 & 0 \\ 0 & -1 & 0 \\ 0 & 0 & 1 \end{pmatrix} \notin \{I,-I\}$.

Proof: From Lemmas a and b we have

$\eta'\sigma' = \begin{pmatrix} 0 & je^{i\phi}(-1)^m e^{i\phi} & 0 \\ \pm ie^{-i\phi} & 0 & 0 \\ 0 & 0 & (\pm 1)i^m \end{pmatrix}$

and a routine (if somewhat tedious) calculation proves the lemma
and hence (†) .

D. SO(n) splits

We showed that the normalizer in SO(3) split, but that was eas
because we knew the Weyl group was just $\mathbb{Z}/2$, so we simply had to
find an element of N - T whose square equalled I . (It turned out

that any element of N - T would do.) For n > 3 we will have

bigger Weyl groups. For notational simplicity we will work with

SO(5) first and then consider the general case.

Let

$$\tau(\theta_1,\theta_2) = \begin{pmatrix} \cos\theta_1 & \sin\theta_1 & 0 & 0 & 0 \\ -\sin\theta_1 & \cos\theta_1 & 0 & 0 & 0 \\ 0 & 0 & \cos\theta_2 & \sin\theta_2 & 0 \\ 0 & 0 & -\sin\theta_2 & \cos\theta_2 & 0 \\ 0 & 0 & 0 & 0 & 0 \end{pmatrix}$$

be an element in our maximal torus. Let P_1,P_2 be the two 2-planes

in R^5 spanned by coordinates 1,2 and 3,4 respectively. Let R

denote the line of the 5th coordinate.

Proposition 4: P_1,P_2,R are the only subspaces stable for each

$\tau(\theta_1,\theta_2)$ in T . If N is the normalizer of T and $A \in N$, then

$$RA = R \quad \text{and}$$

$$P_1A \in \{P_1,P_2\} \ , \ P_2A \in \{P_1,P_2\} \quad .$$

Proof: The first statement is clear. Now RA is some line,

say RA = R' . Since $A \in N$ for every $\tau \in T$ we have $A\tau A^{-1} \in T$,

so that $RA\tau A^{-1} = R$. Then

$$R'\tau A^{-1} = R \quad \text{or} \quad R'\tau = RA = R' \ ,$$

proving that R' = R .

Now P_1A is some 2-plane, say $P_1A = P$. For any $\tau \in T$ we

have

$$P_1 A_T A^{-1} = P_1 .$$

So $P_T = P_1 A = P$. Thus $P \in \{P_1, P_2\}$. Of course the same proof works

for P_2 .

Consider two specific elements of N .

$$A = \begin{pmatrix} 1 & 0 & 0 & 0 & 0 \\ 0 & 1 & 0 & 0 & 0 \\ 0 & 0 & 0 & 1 & 0 \\ 0 & 0 & 1 & 0 & 0 \\ 0 & 0 & 0 & 0 & -1 \end{pmatrix} \qquad B = \begin{pmatrix} 0 & 0 & 0 & 1 & 0 \\ 0 & 0 & 1 & 0 & 0 \\ 0 & 1 & 0 & 0 & 0 \\ 1 & 0 & 0 & 0 & 0 \\ 0 & 0 & 0 & 0 & 1 \end{pmatrix} .$$

We have $A^{-1} = A$ and we easily find that

$$A_T(\theta_1, \theta_2) A = \tau(\theta_1, -\theta_2) .$$

Also $B = B^{-1}$ and

$$B_T(\theta_1, \theta_2) B = \tau(-\theta_2, -\theta_1) .$$

If we think of these as actions on the tangent space $\mathscr{L}(T)$ of

T (i.e., take differentials) we see that these are reflections

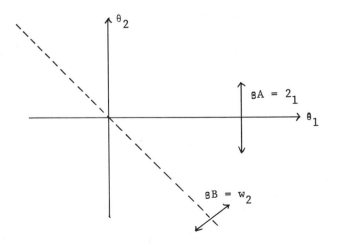

We have $T \overset{\alpha}{\to} N \overset{\beta}{\to} W$. Let $w_1 = \beta(A)$ and $w_2 = \beta(B)$. It is not hard to see that w_1 and w_2 generate a subgroup W' of W with W' containing all permutations and sign changes of $\{\theta_1, \theta_2\}$.

Recall that no two different elements of W have the same action on T (Proposition 2), so if W' contains all possible isomorphisms of T , it will follow that $W' = W$.

Proposition 5: The isomorphisms of $T = S^1 \times S^1$ are in 1-1 correspondence with the isomorphisms of $\mathbb{Z} \times \mathbb{Z}$.

Proof: We can represent T as a quotient group $T = \dfrac{R \times R}{\mathbb{Z} \times \mathbb{Z}}$. Thus any linear isomorphism of $R \times R$ which sends $\mathbb{Z} \times \mathbb{Z}$ onto $\mathbb{Z} \times \mathbb{Z}$ gives an isomorphism of T . But any isomorphism of $\mathbb{Z} \times \mathbb{Z}$ is extended easily to such a linear isomorphism of $R \times R$. Conversely, any isomorphism of T has a differential which is a isomorphism of $R \times R$ sending $\mathbb{Z} \times \mathbb{Z}$ onto $\mathbb{Z} \times \mathbb{Z}$.

Now isomorphisms of $\mathbb{Z} \times \mathbb{Z}$ are easy to determine. They correspond 1-1 with the set of eight permutations-and-sign-changes on $\{(1,0),(0,1)\}$. So the Weyl group of $SO(5)$ is the set of permutations-and-sign-changes on $\{\theta_1, \theta_2\}$. It is generated by the two reflections w_1, w_2 and we have $w_1^2 = 1$, $w_2^2 = 1$ and $(w_1 w_2)^4 = 1$. (This last since $w_1 w_2$ is a rotation of $\frac{\pi}{2}$.)

We have now that $SO(5)$ splits. For if we let $\gamma(1) = I$, $\gamma(w_1) = A$, $\gamma(w_2) = B$ we see that it only remains to check that

$$(AB)^4 = I \ .$$

Well,

$$AB = \begin{pmatrix} 0 & 0 & 0 & 1 & 0 \\ 0 & 0 & 1 & 0 & 0 \\ 1 & 0 & 0 & 0 & 0 \\ 0 & 1 & 0 & 0 & 0 \\ 0 & 0 & 0 & 0 & -1 \end{pmatrix} \quad .$$

Then

$$(AB)^2 = \begin{pmatrix} 0 & 1 & 0 & 0 & 0 \\ 1 & 0 & 0 & 0 & 0 \\ 0 & 0 & 0 & 1 & 0 \\ 0 & 0 & 1 & 0 & 0 \\ 0 & 0 & 0 & 0 & 1 \end{pmatrix}$$

and so

$$(AB)^4 = \begin{pmatrix} 1 & 0 & 0 & 0 & 0 \\ 0 & 1 & 0 & 0 & 0 \\ 0 & 0 & 1 & 0 & 0 \\ 0 & 0 & 0 & 1 & 0 \\ 0 & 0 & 0 & 0 & 1 \end{pmatrix} = I \quad .$$

We have noticed before that the groups $SO(\text{odd})$ and $SO(\text{even})$ are somewhat different. Let us analyze $SO(4)$ just as we have $SO(5)$. Let $C = \begin{pmatrix} 0 & 1 & 0 & 0 \\ 1 & 0 & 0 & 0 \\ 0 & 0 & 0 & 1 \\ 0 & 0 & 1 & 0 \end{pmatrix}$. Then $C^{-1} = C$ and

$$C\tau(\theta_1, \theta_2)C = \tau(-\theta_1, -\theta_2) \quad .$$

Let $B = \begin{pmatrix} 0 & 0 & 0 & 1 \\ 0 & 0 & 1 & 0 \\ 0 & 1 & 0 & 0 \\ 1 & 0 & 0 & 0 \end{pmatrix}$. Then $B \in SO(4)$, $B^{-1} = B$ and

$$B\tau(\theta_1, \theta_2)B = \tau(-\theta_2, -\theta_1) \quad .$$

So $BC\tau(\theta_1,\theta_2)CB = \tau(\theta_2,\theta_1)$. So we can permute θ_1 and θ_2 and
we can change both signs. We claim that no element of W can
change just one sign. Any $D \in N$ must be of the form

or of the form

(see Proposition 4). And each shaded area is an orthogonal 2×2
matrix. If we had a $D \in N$ such that $D\tau(\theta_1,\theta_2)^tD$ changed just
one sign but also permuted θ_1,θ_2 we could combine with B to get
one changing one sign but not permuting. Such a matrix cannot have

the form . But if $D = \begin{pmatrix} \sigma & 0 \\ 0 & \rho \end{pmatrix}$ with σ,ρ being 2×2

orthogonal matrices and $D\tau(\theta_1,\theta_2)^tD = \tau(-\theta_1,\theta_2)$ we easily calculate

that $\rho R(\theta_2)^t\rho = R(\theta_2)$ $(R(\theta) = \begin{pmatrix} \cos\theta & \sin\theta \\ -\sin\theta & \cos\theta \end{pmatrix})$. This implies that

ρ commutes with all rotations. Thus ρ is a rotation, so
$\det \rho = +1$. But then σ is a rotation and we cannot have

$$\sigma R(\theta_1)^t\sigma = R(-\theta_1)$$

as required.

Otherwise the homomorphic cross section proof goes just as for
$SO(5)$. The elements $w_1 = \beta(C)$ and $w_2 = \beta(B)$ generate W and
$\gamma(w_1) = C$, $\gamma(w_2) = B$ generates a homomorphic cross section. The
general case is more difficult only because of notation. The Weyl
group $SO(n)$ is generated by all permutations of $\{\theta_1,\ldots,\theta_n\}$ and
all even sign changes.

E. Exercises

1. Show that an isomorphism of Lie groups induces an isomorphism of their Weyl groups.

2. Prove (+) for $\dfrac{Sp(n)}{center}$ ($n \geq 3$) (i.e., generalize the proof given for $\dfrac{Sp(3)}{center}$) .

3. Show that reflection in L is given by

$$r = \begin{pmatrix} \cos 2\phi & \sin 2\phi \\ \sin 2\phi & -\cos 2\phi \end{pmatrix} .$$

4. Show that, with r as in 3 ,

$$r \begin{pmatrix} \cos \theta & \sin \theta \\ -\sin \theta & \cos \theta \end{pmatrix} r = \begin{pmatrix} \cos \theta & -\sin \theta \\ \sin \theta & \cos \theta \end{pmatrix} .$$

5. Show that the only isomorphism of $S^1 = \{e^{i\theta}\}$ which leaves i fixed is the identity. Show that there is exactly one isomorphism of S^1 which sends i to -i .

Chapter 12
Lie Groups

A. Differentiable manifolds

   If U is an open set in R^n and we have a function $f : U \to R^m$,
we say that f is <u>smooth</u> (or C^∞) if all higher partial derivatives
of f exist and are continuous. The composition of smooth functions
is smooth. In the case $m = n$, if $f : U \to R^n$ is one-to-one onto
$f(U)$, with $f(U)$ open in R^n , and both f and f^{-1} are smooth
then f is a <u>diffeomorphism</u> (from U to $f(U)$) .

   Let M be a n-manifold (§E, Chapter VI). By definition, for any
$p \in M$ we have U open in R^n and a homeomorphism

$$\phi : U \to M$$

into M with $\phi(U)$ being an open neighborhood of p . Such a pair
(U, ϕ) will be called a <u>chart</u>. A collection of charts such that the
$\phi(U)$'s cover M will be called an <u>atlas</u>.

   Two charts $\phi : U \to M$ and $\psi : V \to M$ are said to <u>overlap</u> <u>smoothly</u>

if (either $\phi(U) \cap \phi(V) = \emptyset$, or) $\psi^{-1} \circ \phi$ is a diffeomorphism.

Definition: M is a <u>differentiable manifold</u> if it has an atlas of smoothly overlapping charts.

Given such an atlas we can <u>maximalize</u> it. We simply add in all charts which overlap smoothly with the given ones in the atlas. Since compositions of diffeomorphisms are again diffeomorphisms we see that: If two new charts overlap smoothly with the given charts in the atlas, then they overlap smoothly with each other. We call a maximal atlas a <u>differentiable structure</u>.

Suppose we have differentiable manifolds M and N and a function $f : U \to N$ with U an open set in M . What should it mean for f to be smooth? For $x \in U$ choose charts $\phi : V \to M$ with $\phi(V)$ an open neighborhood of x and $\psi : W \to N$ with $\psi(W)$ an open neighborhood of f(x) . We say f is <u>smooth at</u> x if $\psi^{-1} \circ f \circ \phi$ is smooth at $\phi^{-1}(x)$. It is easy to check that this definition is independent of the choices of charts (within the maximal atlases).

A smooth one-to-one map $f : M \to N$ with a smooth inverse is a <u>diffeomorphism</u> of the differentiable manifolds. If $M = R = N$ and f is given by $f(x) = x^3$, then f is smooth and one-to-one, but f^{-1} is not smooth (since its derivative fails to exist at 0) so that f is not a diffeomorphism.

B.   <u>Tangent vectors</u>, <u>vector fields</u>

If we have a smooth curve or surface in some R^n , then the concep of a vector tangent to the curve or surface is not hard to define. But our differentiable manifolds are to be thought of as spaces in their own right - not necessarily sitting in some euclidean space. How then do we define "tangent vector?" The idea is that if we have a vector in R^n we can use it to differentiate functions - essentially taking directional derivatives. So we will call a tangent vector a thing which differentiates functions.

Let M be a differentiable m-manifold and $p \in M$. We set

$$A(p) = \{(U,f) \mid p \in U \text{ open, } f : U \to R \text{ smooth}\} .$$

We used $A(p)$ to denote this set of pairs (U,f) since we are going to make it into an algebra. It is routine to verify that if we define

$$(U,f) + (V,g) = (U \cap V, f + g) \quad \text{and}$$

$$r(U,f) = (U, rf) ,$$

that these operations make $A(p)$ into a real vector space. Then we define

$$(U,f)(V,g) = (U \cap V, fg)$$

and $A(p)$ becomes an algebra.

Definition: A tangent vector η to M at p is a linear map $\eta : A(p) \to R$ such that:

(i) if $f = g$ on some neighborhood of p, then $\eta(f) = \eta(g)$.

(ii) $\eta(fg) = f(p)\eta(g) + \eta(f)g(p)$.

Suppose f is a constant function, $f(x) = r$ for all x in some neighborhood of p . Then $fg = rf$ in that neighborhood, and if η is any tangent vector we have $\eta(rg) = r\eta(g)$, since η is linear. On the other hand, by (ii) we have

$$\eta(rg) = \eta(fg) = f(p)\eta(g) + \eta(f)g(p)$$

$$= r\eta(g) + \eta(f)g(p) \quad .$$

So we see that $\eta(f)g(p) = 0$ for all g . Thus $\eta(f) = 0$. So a tangent vector sends any constant function to zero. Also if we have two functions f,g such that $f(p) = 0 = g(p)$, then for any tangent vector η to M at p we have $\eta(fg) = 0$ (by (ii)) .

Any chart $\phi : U \to M$ (with $p \in \phi(U)$) in the differentiable struc-ture gives some tangent vectors as follows. If f is in $A(p)$, then $f \circ \phi$ is a map of $U \subset R^n$ into R and we let

$$\partial_i(p)(f)$$

be the ith partial derivative of $f \circ \phi$ at $\phi^{-1}(p)$. Thus $\partial_i(p) : A(p) \to R$ is a tangent vector. So the chart (U,ϕ) gives n tangent vectors $\partial_1(p)$,..., $\partial_n(p)$ at p .

Let T_pM be the set of all tangent vectors to M at p . If we define operations by

$$(\xi+\eta)(f) = \xi(f) + \eta(f)$$

$$(r\xi)(f) = r\xi(f) \quad ,$$

then it is routine to verify that T_pM becomes a real vector space.

Proposition 1: $\dim T_pM = n$ $(= \dim M)$.

Proof: We have seen that if we take a chart $\phi : U \to M$ in the differentiable structure with $p \in \phi(U)$, we get n tangent vectors $\partial_1(p)$,..., $\partial_n(p)$ in T_pM . We will show that these are a basis for T_pM .

We may assume $\phi(0) = p$. Then for any $f \in A(p)$ we can write

$$f(x) = f(p) + \sum_{i=1}^{n} \partial_i(p)(f)x_i + \sum_{i,j=1}^{n} \phi_{ij}(x)x_ix_j$$

for suitable smooth functions ϕ_{ij} . This is just the multivariate version of the following result about functions of one variable. Let f be a smooth real-valued function defined on some neighborhood of 0 in R . We note that

$$\psi(x) = \frac{\frac{f(x)-f(0)}{x} - f'(0)}{x}$$

is smooth (and defined for $x \neq 0$) . Solving for $f(x)$ gives

$$f(x) = f(0) + f'(0)x + \psi(x)x^2 .$$

Now let η be any element of T_pM and apply η to f . η is linear, so $\eta(f(p)) = 0$ since $f(p)$ is constant, and $\eta(\phi_{ij}(x)x_ix_j)$ $= 0$ since both $\phi_{ij}(x)x_i$ and x_j vanish at p . Thus

$$\eta(f) = \eta(x_1)\partial_1(p)(f) +...+ \eta(x_n)\partial_n(p)(f) ,$$

and we conclude that

$$\eta = \eta(x_1)\partial_1(p) +...+ \eta(x_n)\partial_n(p) ,$$

showing that $\{\partial_1(p)$,..., $\partial_n(p)\}$ spans T_pM .

Now these tangent vectors are linearly independent, for if

$$c_1 \partial_1(p) + \ldots + c_n \partial_n(p)$$

is the zero tangent vector and we apply it to the function x_i we get $c_i = 0$.

q.e.d.

Definition: A vector field X on an open set W in M is an assignment of $X_p \in T_p M$ for each $p \in W$.

We are interested in vector fields which are continuous in the sen that if points p and q are close then X_p and X_q are close. But since X_p and X_q are in different vector spaces, this looks difficult to formulate. Actually there is a neat way of even defining a smooth vector field.

Let X be a vector field on W and let $f : W \to R$ be smooth. We get a new function

$$Xf : W \to R$$

defined by $(Xf)(p) = X_p(f)$. This makes sense because $f \in A(p)$ for each $p \in W$.

Definition: The vector field X is smooth if for each smooth f we have Xf smooth.

It is easy to verify that if X and Y are smooth vector fields then so is $X + Y$,

$$(X+Y)(f) = X(f) + Y(f) \quad .$$

Also if X is a smooth vector field and $r \in R$, then rX is smooth $(rX(f) = r(X(f)))$. Thus the smooth vector fields form a vector space

For a coordinate chart $\phi : U \to M$ in the differentiable structure we have smooth vector fields $\partial_1, \ldots, \partial_n$. So if $\sigma_1, \ldots, \sigma_n$ are

smooth real-valued functions on U we have

$$\sigma_1 \partial_1 + \ldots + \sigma_n \partial_n$$

as a smooth vector field. Conversely, if X is smooth on U , we

have

$$X = \rho_1 \partial_1 + \ldots + \rho_n \partial_n \quad ,$$

and the functions ρ_i are smooth because ρ_i is $X(x_i)$, the

function obtained by applying the smooth vector field X to the

smooth function x_i .

If we have two smooth vector fields X and Y we can get a real-

valued operator on real-valued smooth functions as follows:

$$f \mapsto X_p(Yf) \quad .$$

Since Y and f are smooth, so is Yf and therefore X_p assigns

the real number $X_p(Yf)$ to this function. But this operator from

$A(p)$ to $\mathbb{R}$ need not be a tangent vector. It is linear and depends

only on f near p (condition (i)), but it may fail to satisfy con-

dition (ii) in the definition of a tangent vector. Indeed, if $X = \partial_i$

and $Y = \partial_j$, then (ii) is false because the operation is just the

mixed second partial derivative.

Proposition 2: For smooth vector fields, X, Y the operator

$$f \mapsto X_p(Yf) - Y_p(Xf)$$

is a tangent vector. ("The mixed second partials cancel out") .

Proof: Let X be a smooth field and f,g be smooth functions. We assert that

(+) $$X(fg) = (Xf)g + f(Xg) \quad .$$

At a point p the left hand side is

$$X_p(fg) = X_p(f)g(p) + f(p)X_p(g)$$

since X_p satisfies (ii), and this is just the right hand side evaluated at p . Of course, we have a similar formula for Y(fg) .

Then

$$X_p(Y(fg)) = X_p((Yf)g + f(Yg))$$

$$= X_p(Yf)g(p) + (Yf)(p)X_p(g) + X_p(f)(Yg)(p) + f(p)X_p(Yg) \quad ,$$

and similarly

$$Y_p(X(fg)) = Y_p(Xf)g(p) + (Xf)(p)Y_p(g) + Y_p(f)(Xg)(p) + f(p)Y_p(Xg) \quad .$$

Thus

$$(X_pY - Y_pX)(fg) = (X_pY - Y_pX)(f)g(p) + f(p)(X_pY - Y_pX)(g) \quad ,$$

proving that condition (ii) is satisfied. The linearity and condition (i) are true since they are true for both terms. So

$$X_pY - Y_pX \quad \text{is a tangent vector at} \quad p \quad .$$

Let $[X,Y] = XY - YX$ be the vector field defined by

$$[X,Y]_p = X_pY - Y_pX \quad .$$

Proposition 3: The set $\mathfrak{X}(W)$ of smooth vector fields on an open set W in a differentiable manifold M forms a Lie algebra under [,] .

Proof: We have seen that $\mathfrak{X}(W)$ is a vector space. From the definition of [X,Y] it is obvious that [Y,X] = -[X,Y] and it is easy to show that this multiplication distributes over addition ([X+Y,Z] = [X,Z] + [Y,Z]) . Proof of the Jacobi identity is formally just the same as for matrix multiplication with [A,B] = AB - BA .

This Lie algebra $\mathfrak{X}(W)$ is usually infinite dimensional, but when our differentiable manifold is a Lie group we will get an important finite-dimensional subalgebra. (See section C).

Let M,N be differentiable manifolds and

$$M \xrightarrow{\psi} N$$

be a smooth map. If $\psi(p) = q$ we get a map

$$T_p M \xrightarrow{d\psi} T_q N$$

called the differential of ψ as follows:

If $\xi \in T_p M$ and $f \in A(q)$ we set

$(\star)$ $\qquad\qquad\qquad\qquad d\psi\,\xi(f) = \xi(f \circ \psi)$.

Proposition 4: $(\star)$ defines a map of $T_p M$ into $T_q N$ and it is a linear map.

Proof: We need to show that $d\psi\,\xi$ is a linear derivation from A(q) to R and assigns the same real number to two functions which agree on any neighborhood of q . This final condition is inherited from the same property for ξ and linearity is easy to prove. So suppose $f, g \in A(q)$. We have

$$d_\psi \xi(fg) = \xi(fg \circ \psi) = \xi((f \circ \psi)(g \circ \psi))$$

$$= (f \circ \psi)(p)\xi(g \circ \psi) + \xi(f \circ \psi)(g \circ \psi)(p)$$

$$= f(q)d_\psi \xi(g) + d_\psi \xi(f)g(q) \quad ,$$

showing that $d_\psi \xi$ is a derivation.

Finally $d\psi : T_p M \rightarrow T_q N$ is linear. For we have

$$d\psi(a\xi + b\eta)(f) = (a\xi + b\eta)(f \circ \psi) = a\xi(f \circ \psi) + b\eta(f \circ \psi) = (ad_\psi \xi + bd_\psi \eta)(f)$$

C. Lie groups

Let G be a differentiable n-manifold which is also a group and the operations

$$G \times G \rightarrow G \qquad G \rightarrow G$$

$$(a,b) \mapsto ab \qquad a \mapsto a^{-1}$$

are smooth functions. Then G is called a Lie group.

Let G be a Lie group with identity element e , and suppose X_e is a tangent vector at e $(X_e \in T_e G)$. Then we can get a vector field defined on all of G as follows. For any $g \in G$ let $L_g : G \rightarrow G$ be the diffeomorphism given by $L_g(x) = gx$ for each $x \in G$. This is called left-translation by g . We set

$$(\dagger) \qquad X_g = dL_g X_e$$

$(dL_g : T_e G \to T_g G)$. Such vector fields are called <u>left-invariant</u>; i.e.
a vector field X on G is left-invariant if it satisfies (+) .

Proposition 5: <u>If</u> X,Y <u>are left-invariant vector fields on</u> G ,
<u>so is</u> [X,Y] .

Proof: Let $g \in G$ and $f \in A(g)$ and we calculate

$$dL_g [X,Y]_e (f) = [X,Y]_e (f \circ L_g)$$

$$= X_e (Y(f \circ L_g)) - Y_e (X(f \circ L_g))$$

$$= dL_g X_e (Yf) - dL_g Y_e (Xf)$$

$$= X_g (Yf) - Y_g (Xf)$$

$$= [X,Y]_g (f) \ , \ \text{proving (+)} \ .$$

Now one can see easily that X,Y left invariant implies X+Y
is also and so is rX $(r \in R)$. Thus the set of left-invariant
fields on G becomes a subalgebra of the Lie algebra of all smooth
vector fields. Since left-invariant vector fields correspond one-to-
one with elements of $T_e G$, this Lie algebra is n-dimensional. We
denote it by $\mathcal{L}(G)$ and call it the <u>Lie algebra of</u> G . It is convenient to use the language of categories and functors to discuss
passing from G to $\mathcal{L}(G)$.

Categories and Functors

A category consists of <u>objects</u> A,B,C,... , and for each pair
,B of objects a set Hom(A,B) called the <u>morphisms</u> from A to B .
ach Hom(A,A) is required to contain the identity morphism i_A ,

and there must be a <u>law</u> <u>of</u> <u>composition</u> so that $\alpha \in \text{Hom}(A,B)$ and $\beta \in \text{Hom}(B,C)$ give a unique $\beta \circ \alpha \in \text{Hom}(A,C)$.

<u>Examples</u>:

(i) objects - sets

morphisms - functions

composition is ordinary composition of functions

(ii) objects - groups

morphisms - homomorphisms

ordinary composition

(iii) objects - vector spaces

morphisms - linear maps

ordinary composition

(iv) objects - differentiable manifolds

morphisms - smooth maps

<u>Definition</u>: $\gamma \in \text{Hom}(A,B)$ is an <u>isomorphism</u> if there exists $\delta \in \text{Hom}(B,A)$ such that

$$\delta \circ \gamma = i_A \quad \text{and} \quad \gamma \circ \delta = i_B .$$

In (i) an isomorphism is a one-to-one function, in (ii) it is a group isomorphism, in (iii) it is a linear isomorphism and in (iv) it is a diffeomorphism.

If C_1 and C_2 are categories a <u>functor</u>

$$F : C_1 \to C_2$$

must send objects to objects and morphisms to morphisms such that:

we have either

$$
\begin{array}{ccc}
A & \longrightarrow & F(A) \\
\alpha \downarrow & & \downarrow F(\alpha) \\
\text{(I)} \quad\quad B & \longrightarrow & F(B) \quad\quad \text{with} \;\; F(\beta \circ \alpha) = F(\beta) \circ F(\alpha) \\
\beta \downarrow & & \downarrow F(\beta) \\
C & \longrightarrow & F(C)
\end{array}
$$

or

$$
\begin{array}{ccc}
A & \longrightarrow & F(A) \\
\alpha \downarrow & & \uparrow F(\alpha) \\
\text{(II)} \quad\quad B & \longrightarrow & F(B) \quad\quad \text{with} \;\; F(\beta \circ \alpha) = F(\alpha) \circ F(\beta) \\
\beta \downarrow & & \uparrow F(\beta) \\
C & \longrightarrow & F(C)
\end{array}
$$

In case (I) F is called a <u>covariant</u> functor and in case (II) it is a <u>contravariant</u> functor.

If C_1 and C_2 are both example (ii) above, then the functor which assigns to V its dual $V*$ gives a contravariant functor.

We have a category with Lie groups as objects and smooth homomorphisms as morphisms and we have a category of Lie algebras and Lie algebra homomorphisms.

<u>Proposition</u> 6: <u>The</u> <u>assignment</u> $G \to \mathfrak{L}(G)$ <u>gives</u> <u>a</u> <u>covariant</u> <u>functor</u> <u>from</u> <u>the</u> <u>Lie</u> <u>group</u> <u>category</u> <u>to</u> <u>the</u> <u>Lie</u> <u>algebra</u> <u>category.</u>

<u>Proof</u>: We must first say how a homomorphism of Lie groups induces a homomorphism of their Lie algebras. Let

$$G \xrightarrow{\phi} H$$

be a homomorphism of Lie groups. We get a linear map

$$T_eG \xrightarrow{d\phi} T_eH$$

by taking the differential of ϕ . Since these can be identified with $\mathfrak{L}(G)$ and $\mathfrak{L}(H)$ we have a linear map of the Lie algebras. We must just check that it is a __homomorphism__ of Lie algebras; i.e. that it preserves the product $[X,Y]$.

Let $X,Y \in \mathfrak{L}(G)$ (i.e. left-invariant vector fields on G) and we calculate

$$d\phi[X,Y]f = [X,Y](f \circ \phi) = X(Y(f \circ \phi)) - Y(X(f \circ \phi))$$

$$= d\phi X(Yf) - d\phi Y(Xf) = [d\phi X, d\phi Y]f \ \ .$$

Since this holds for all f , $d\phi[X,Y] = [d\phi X, d\phi Y]$ as required.

The functor $\mathfrak{L}$ does not map objects in a one-to-one manner -- two nonisomorphic Lie groups can have isomorphic Lie algebras. In fact, if $\phi : G \to H$ is a Lie group homomorphism which is a diffeomorphism on some neighborhood of e in G , the Lie algebra homomorphism $d\phi : \mathfrak{L}(G) \to \mathfrak{L}(H)$ will be an isomorphism of Lie algebras. Recall the two-to-one homomorphism ρ of $Sp(1)$ onto $SO(3)$. We see that $d\rho$ is an isomorphism of their Lie algebras (we proved these are isomorphic in Exercise 5 Chapter IV).

It turns out, however, that every real finite-dimensional Lie algebra is the Lie algebra of some Lie group. Indeed there is the following theorem (which we are not prepared to prove here - see page 133 of Hochschild, The Structure of Lie Groups, Holden-Day, 1965).

Theorem: Let $\mathcal{L}$ be a finite dimensional real Lie algebra. Let G be the group of self-isomorphisms of $\mathcal{L}$. Then $\mathcal{L}(G) \cong \mathcal{L}$.

We conclude this section with some remarks about subgroups and subalgebras. We would like somehow to have a subgroup H of a Lie group G have its Lie algebra $\mathcal{L}(H)$ be a subalgebra of the Lie algebra $\mathcal{L}(G)$. The catch is that H may not be a Lie group at all. The simple classical example is: $G = S^1 \times S^1$ (a 2-torus). $G = \exp(R^2)$ where $\exp(x,y) = (e^{2\pi ix}, e^{2\pi iy})$. Let L be a line through $(0,0)$ in R^2 making angle θ with the x-axis where θ is chosen so that $\frac{2\pi}{\theta}$ is irrational. Let $H = \exp(L)$. This is a subgroup of R^2 , but is not a Lie group since it is not a manifold. It is dense in G (every point of G is a limit point of H). If, on the other hand, $\frac{2\pi}{\theta}$ is rational then H will be a circle subgroup of G . The result (which we will not prove here) is that a closed subgroup of a Lie group is itself a Lie group. For G and H as first described (H bad) we have that $\mathcal{L}(G) = R^2$ is the trivial Lie algebra (always $[X,Y] = 0$) and thus L is a Lie subalgebra of $\mathcal{L}(G)$. $\exp(L)$ is a subgroup H of G, but the subalgebra L cannot be $\mathcal{L}(H)$ because $\mathcal{L}(H)$ does not exist.

D. Connected groups

Proposition 7: Let X be a pathwise connected space and let Y be a subset of X which is both open and closed. Then if $Y \neq \phi$, Y = X .

Proof: We assume Y is not empty and choose $y \in Y$. For any $x \in X$ take a path $\rho : [0,1] \to X$ from y to x.

Let C be the points $t \in [0,1]$ which map into Y. We have $0 \in C$ (since $\rho(0) = y \in Y$), and we will show $1 \in C$ so that $x \in Y$.

Now C is closed. For any limit point of C maps to a limit point of Y (by continuity of ρ). This limit point of Y belongs to Y so, by definition of C, the limit point of C belongs to C

If $1 \notin C$ let t_0 be the least upper bound of C. Since C is closed $t_0 \in C$, and thus $\rho(t_0) \in Y$. Since Y is open and ρ is continuous, some interval $(t_0 - \epsilon, t_0 + \epsilon)$ maps into Y, contradicting that t_0 is the least upper bound of C. Thus $1 \in C$ and $x \in Y$.

Proposition 8: If G is a pathwise connected Lie group and H is a subgroup which contains an open neighborhood U of e in G, then H = G.

Proof: Let

$$U^2 = \{xy \mid x \in U, y \in U\}$$

$$U^3 = \{xyz \mid x \in U, y \in U, z \in U\}, \quad \text{etc.}$$

Since H is a subgroup we see that

$$W = U \cup U^2 \cup U^3 \ldots$$

lies in H. As a union of open sets, W is an open set. But W i also closed, for suppose x is a limit point of W. Then the open set xU contains some

$$u_1 u_2 \cdots u_m \in W \quad (\text{each } u_i \in U) \quad .$$

That is, there is $u \in U$ such that

$$xu = u_1 u_2 \cdots u_m \; ,$$

but then $x = u_1 u_2 \cdots u_m u^{-1} \in W$.

By Proposition 7 we see that W being both open and closed and nonempty implies that $W = G$. Since $W \subset H$ we have $H = G$.

Corollary: Let $f : K \to G$ be a homomorphism of Lie groups with G pathwise connected. If $f(K)$ contains some open neighborhood of e in G , then f is surjective.

If G is a Lie group which is not connected, let G_0 denote all $x \in G$ which can be connected to the identity element e by a path in G . We call G_0 the identity component of G .

Proposition 9: The identity component G_0 of G is a pathwise connected subgroup of G which is both open and closed in G .

Proof: G_0 is pathwise connected, for if $x, y \in G_0$ we can choose paths ρ, σ from e to x, y and then $\rho^{-1} \sigma$ is a path from x to y and it lies in G_0 since each $\rho(t)$ and $\sigma(t)$ belong to G_0 .

G_0 is a subgroup because it contains e and if $x, y \in G_0$ with paths ρ, σ then the path

$$(\rho\sigma)(t) = \rho(t)\sigma(t)$$

is a path in G from e to xy . Thus $xy \in G_0$. Similarly, if ρ is a path from e to x , then

$$\tau(t) = (\rho(t))^{-1}$$

is a path from e to x^{-1} .

G_0 is open. For if $x \in G_0$ we can take an open neighborhood U of x in G homeomorphic to an open ball in R^n (G is an n-manifold). Since x can be connected by a path ρ from e and each point in U can be connected by a path from x in U , we see that $U \subset G_0$.

G_0 is closed. For if x is a limit point of G_0 we can take U as above. Some point y in U is in G_0 and so we have a path σ from e to y . Since we have in U a path from y to x we see that $x \in G_0$.

We are going to need an <u>exponential</u> <u>map</u> from $T_e G$ to G , and we cannot use the one from Chapter IV because the elements of $T_e G$ and G cannot be assumed to be matrices. If $\gamma : (-\epsilon, \epsilon) \to G$ is a smooth curve with $\gamma(0) = e$ we define the <u>derivative</u> $\gamma'(0)$ of γ at e to be

$$\gamma'(0) = d\gamma(\xi)$$

where ξ is the unit vector 1 in R $(= T_0(R))$. (In the exercise one shows that for matrix groups this agrees with our definition in Chapter II) .

We will need the following uniqueness theorem from differential-equation theory (see Milnor, Morse Theory, Annals of Math Studies #51, 1963, Lemma 2.4).

<u>Theorem</u>: <u>Given</u> <u>any</u> $\eta \in T_e G$, <u>there</u> <u>exists</u> <u>exactly</u> <u>one</u> <u>one-parameter</u> <u>subgroup</u> γ <u>in</u> G <u>such</u> <u>that</u>

$$\gamma'(0) = \eta \quad . \quad \text{(see Chapter IV)} \quad .$$

<u>Definition</u>:   Given any $\eta \in T_e G$,   take γ to be the one-parameter subgroup in G such that $\gamma'(0) = \eta$ and set

$$\exp(\eta) = \gamma(1) \quad .$$

Let η and γ be as in this definition.   Set

$$\rho(t) = \exp(t\eta) \quad .$$

We claim that this curve in G is just the one-parameter subgroup γ .
To see this hold t fixed and consider the curve $\sigma(u) = \gamma(tu)$.
Then σ is easily seen to be a one-parameter subgroup and

$$\sigma'(0) = t\gamma'(0) = t\eta \quad .$$

Thus, by definition of $\exp$, $\exp(t\eta) = \sigma(1) = \gamma(1) = \gamma(t)$.   So we
have proved

<u>Proposition</u> 10: $t \mapsto \exp(t\eta)$ <u>is the unique one-parameter sub-</u><u>group whose derivative at</u> 0 <u>is</u> η .

This implies that   exp   when   restricted to a one-dimensional
subspace of $T_e G$ is a homomorphism into G .

For matrix groups we saw that the exponential map was a diffeomorphism of some neighborhood of I in the group (having   log   as
inverse).   This is still true in our more general setting and is based
on the

<u>Inverse Function Theorem</u>:   Let M,N <u>be differentiable n-manifold</u>
<u>and</u> $\phi : M \to N$ <u>be a smooth map.   If</u> $d\phi : T_p M \to T_{\phi(p)} N$ <u>is an</u>

isomorphism, then ϕ is a diffeomorphism of some open neighborhood of p .

(A proof of this can be found in most good advanced calculus books).

We now apply this theorem to

$$\exp : T_e G \to G \quad .$$

We have $d(\exp) : T_e G \to T_e G$. Let $\eta \in T_e G$ and let γ be the unique one-parameter subgroup such that $\gamma'(0) = \eta$. Then (as we saw in the proof of Proposition 10) $\exp(t\eta) = \gamma(t)$ so that γ is the image under exp of the line $t \mapsto t\eta$. So $d(\exp)$ sends the tangent vector to this line (namely η) to $\gamma'(0)$ (which is also η) . Thus

Proposition 11: $d(\exp) : T_e G \to T_e G$ is the identity map .

Corollary: If G is a pathwise connected Lie group, then $\exp : T_e G \to T_e G$ is surjective. If G is not pathwise connected, $\exp T_e G$ is the identity component G_0 of G .

E. Abelian groups

Just as for abelian matrix groups we have the following result (and the proof is the same).

Proposition 12: Let G be an abelian Lie group. If γ, ρ are one-parameter subgroups, then so is their product.

Again, just as for abelian matrix groups we have

Proposition 13: Let G be an abelian Lie group. Then for any
$X, Y \in T_e G$, we have

$$\exp(X+Y) = \exp(X) \, \exp(Y) \quad .$$

Proof: Let γ, ρ be the one-parameter subgroups satisfying
$\gamma'(0) = X$, $\rho'(0) = Y$. By Proposition 12, $\gamma\rho$ is a one-parameter
subgroup and $(\gamma\rho)'(0) = \gamma'(0)\rho(0) + \gamma(0)\rho'(0) = \gamma'(0) + \rho'(0) = X+Y$.
Thus $\exp(X+Y) = (\gamma\rho)(1) = \gamma(1)\rho(1) = \exp(X) \, \exp(Y)$.

Thus for an abelian Lie group G , exp is a homomorphism from
the additive vector group of $T_e G$ to the group G . Taking note of
the corollary to Proposition 11, we see that if G is pathwise connected, then exp is surjective. Thus a connected abelian group is
a quotient group of the vector group $T_e G$.

Again, just as for matrix groups, exp is a diffeomorphism of
some neighborhood of 0 in $T_e G$ so that the kernel is a discrete
subgroup of the vector group $T_e G$. So just as for matrix groups

Theorem: A connected abelian Lie group G is isomorphic to a
cartesian product of a torus and a vector group. If G is also compact, it is a torus.

Appendix

In Exercise 8 of Chapter VI the concept of <u>limit</u> <u>point</u> is defined and it is proved that C is closed $\Leftrightarrow$ (x lp $C \Rightarrow x \in C$) ; i.e., a set is closed if and only if it contains all of its limit points.

<u>Lemma</u>: <u>If</u> $C \subset R^n$ <u>is bounded</u> <u>and</u> <u>infinite</u> <u>then</u> C <u>has a limit point</u>.

<u>Proof</u>: We must find $x \in R^n$ such that each $B(x, \epsilon) \cap C$ is an infinite set.

Since C is bounded we can find an n - dimensional cube K_1 containing C . Let λ be the length of the sides of K_1 . Divide each edge into two equal parts and cut K_1 into 2^n equal cubes of side length $\frac{\lambda}{2}$. Call those cubes $K_{11}, K_{12}, \ldots, K_{12^n}$. At least one of those must contain infinitely-many points of C . Choose one such and call it K_2 . Subdivide K_2 into 2^n cubes of side $\frac{\lambda}{4}$ and choose one of these, K_3 , containing infinitely-many points of C . Continue this process to get $K_1 \supset K_2 \supset K_3 \supset \cdots$. Clearly $K_1 \cap K_2 \cap K_3 \cap \cdots$ is a single point $x \in R^n$.

Then x lp C . For, consider any $B(x, \epsilon)$. Take m such that $\frac{\lambda}{2^m} < \frac{\epsilon}{2}$. Then

$$K_m \subset B(x, \epsilon)$$

and K_m contains infinitely-many points of C .   q.e.d.

<u>Theorem</u>: (Proposition 5 of Chapter VI) <u>If</u> $C \subset R^n$ <u>is closed</u> <u>and bounded</u> <u>and</u>

$$f: C \to R^m$$

is continuous, then $f(C)$ is closed and bounded.

     Proof: Suppose $f(C)$ is not bounded. Choose $x_1 \in C$ such that $y_1 = f(x_1) \notin B(o,1)$.

Choose $x_2 \in C$ such that $y_2 = f(x_2) \notin B(o,2)$.

$\vdots$

Choose $x_k \in C$ such that $y_k = f(x_k) \notin B(o,k)$. It is easy to prove that $Y = \{y_1, y_2, \ldots\}$ has no limit point and that $X = \{x_1, x_2, \ldots\}$ is an ifinite set. Also $X \subset C$ is bounded. So by the lemma X has some limit point x . Then also x lp C . Since C is closed $x \in C$. But then $y = f(x)$ must be a limit point of $Y = f(X)$ by continuity. Thus $f(C)$ is bounded.

     Suppose $f(C)$ is not closed. Then there exists $y \in R^m$ such that y lp $f(C)$ but $y \notin f(C)$. Choose $x_1 \in C$ such that $y_1 = f(x_1) \in B(y,1)$.

Choose $x_2 \in C$ such that $y_2 = f(x_2) \in B(y,\frac{1}{2})$.

$\vdots$

Choose $x_k \in C$ such that $y_k = f(x_k) \in B(y,\frac{1}{2})$. Clearly y is the only limit point of $Y = \{y_1, \ldots, y_k, \ldots\}$. Just as before $X = \{x_1, \ldots, x_k, \ldots\}$ is infinite and bounded. Let x lp X . Then x lp C so $x \in C$ and $f(x)$ lp $Y = f(X)$. This implies $f(x) = y$, but $f(x) \in f(C)$. So $f(C)$ is closed.   q.e.d.

Index

190

ISBN 0-387-**90462**-X
ISBN 3-540-**90462**-X